ADVANCED TOPICS
IN SCIENCE AND TECHNOLOGY IN CHINA

ADVANCED TOPICS IN SCIENCE AND TECHNOLOGY IN CHINA

Zhejiang University is one of the leading universities in China. In Advanced Topics in Science and Technology in China, Zhejiang University Press and Springer jointly publish monographs by Chinese scholars and professors, as well as invited authors and editors from abroad who are outstanding experts and scholars in their fields. This series will be of interest to researchers, lecturers, and graduate students alike.

Advanced Topics in Science and Technology in China aims to present the latest and most cutting-edge theories, techniques, and methodologies in various research areas in China. It covers all disciplines in the fields of natural science and technology, including but not limited to, computer science, materials science, life sciences, engineering, environmental sciences, mathematics, and physics.

Zhenyao Shen
Junfeng Niu
Ying Wang
Hongyuan Wang
Xin Zhao

Distribution and Transformation of Nutrients and Eutrophication in Large-scale Lakes and Reservoirs

The Three Gorges Reservoir

With 63 figures

ZHEJIANG UNIVERSITY PRESS
浙江大学出版社

Springer

Authors

Prof. Zhenyao Shen
School of Environmental
Beijing Normal University
Beijing 100875, China
E-mail: zyshen@bnu.edu.cn

Prof. Junfeng Niu
School of Environmental
Beijing Normal University
Beijing 100875, China
E-mail: junfengn@bnu.edu.cn

Dr. Ying Wang
School of Environmental
Beijing Normal University
Beijing 100875, China
E-mail:
wangying_21cn@yahoo.com.cn

Dr. Hongyuan Wang
Institute of Agricultural Resources and Regional Planning
Chinese Academy of Agricultural Sciences
Beijing 100081, China
E-mail: wanghy@caas.ac.cn

Dr. Xin Zhao
Water Environment Department
Changjiang River Scientific Research Institute
Wuhan 430010, China
E-mail: zhxlzu@163.com

ISSN 1995-6819 e-ISSN 1995-6827
Advanced Topics in Science and Technology in China

ISBN 978-7-308-10490-6
Zhejiang University Press, Hangzhou

ISBN 978-3-642-34963-8 ISBN 978-3-642-34964-5 (eBook)
Springer Heidelberg New York Dordrecht London

Library of Congress Control Number: 2012951408

Printed on acid-free paper

Springer is part of Springer Science+Business Media (www.springer.com)

Preface

The objectives of this book are to provide a clear description of early eutrophication in large-scale lakes and reservoirs and to present readers with an overview of large-scale lake and reservoir management problems and the tools that can be applied to solve these problems. This book recognizes the need for a description of both the opportunities and limitations inherent in the distribution and transformation law of nutrients in large-scale lakes and reservoirs. The Three Gorges Reservoir, one of the largest dam projects in the world as an example, will draw great concern from the public and government. This book presents some research results of early eutrophication in the Three Gorges Reservoir. Lake management tools are presented in detail, including environmental technological methods, ecotechnological methods and the application of models to assess the best management strategy.

The intent of this book is to present an integrated coverage of hydrodynamics, sediment processes, fate of nutrients and transport, and water quality and eutrophication in large-scale lakes and reservoirs. We hope we have provided a timely book that will be a resource for graduate students, environmental engineers, environmental scientists and ecological chemistry researchers with an interest in the environmental processes, mathematical modeling and fate of nutrients in large-scale lakes and reservoirs.

We would like to take this opportunity to thank all the authors who have offered their contributions and also the financial support of the National Basic Research Program of China (973 Program, Nos. 2003CB415204 and 2010CB429003) that enabled this book to come to fruition.

The authors
Beijing, China
September, 2012

Contents

1

Distribution and Transformation of Nutrients in Large-Scale Lakes and Reservoirs

We present readers with an overview of lake management problems and the tools that can be applied to solve problems. Lake management tools are presented in detail, including environmental technological methods, ecotechnological methods and the application of models to assess the best management strategy.

1.1 Introduction

Nutrients are elements that are the basic atomic building blocks of living tissues. There are 16 elements generally considered as necessary nourishment. C, H, O, N, P, K, S, Ca, Mg are the macro-nutrients and the micro nutrients are like Fe, Mn, B, Zn, Cu, Mo and Cl. However, the atmosphere is the major reservoir of nitrogen on earth. Nitrogen is present in the atmosphere in its elemental form (diatomic N2) and it also has a very strong triple bond which is very hard to break. Even in the aquatic environment, the dominant elements are hydrogen and oxygen. However, a variety of salts are dominated by the cations Na^+, K^+, Mg^{2+}, and Ca^{2+} and the anions Cl^-, SO_4^{2-} and NO_3^- which are essential for aquatic organisms as nourishment. The aquatic environments such as lakes and reservoirs are often referred to as standing waters. Man-made reservoirs, or dams, are purpose built principally to supply water to homes, industry and agriculture or, in some cases, for electrical power generation. Basically, lakes are divided into three trophic categories: oligotrophic, mesotrophic and eutrophic. An oligotrophic lake is a large deep lake with crystal clear waters and a rocky or sandy shoreline. Both planktonic and rooted plant growth are sparse, and the lake can sustain a cold water fishery. A eutrophic lake is typically shallow with a soft and muddy bottom. Rooted plant growth is profuse along the shore and out into the lake.

Water can enter lakes as well as reservoirs from a variety of sources including groundwater, runoff from the watershed, surface waters (like streams and rivers) flowing into the lake, evaporation and direct precipitation into the lake. Flowing water typically undergoes significant changes as the water enters the lake or reservoir, primarily because its velocity reduces: sediments, nutrients and other material carried in the faster-flowing water settle out in the basin, undergoing sedimentation (Ford, 1990). As result, water accumulating huge amounts of nutrients from natural environments is often called eutrophic. Compared to natural lakes, reservoirs tend to be more influenced by nutrients and other substances transported from the surrounding land. Lakes and reservoirs also differ in the amount of phytoplankton and aquatic plants (primary production) that can be supported. Elevated levels of nitrogen and phosphorous from agricultural runoff, and also from fertilizers, liberate the phosphorous and nitrogen limitations that phytoplankton experience and lead to an algal bloom or eutrophication. These blooms might stimulate bacterial growth and reduce dissolved oxygen levels in lakes, which makes aquatic life miserable. Accordingly, the two most noticeable markers of heavy nutrient loading in lakes are an excessive plant growth (eutrophication) and a decreased concentration of dissolved oxygen. However, it is a very slow and natural process; it could be significantly accelerated by human activities that increase the flow of nutrient input in a water body.

Presently, eutrophication is one of the main factors causing rapid growth of micro-organisms and turbid waters in Donghu Lake, China (He et al., 2002). Excessive growth of Eicchornia crassipes and Alternanthera pheloxirodes has been noted in the shallow, eutrophic Donghu Lake. The blooming in terms of biomass and height of the species was noted in the month of November in 1996 and 1998. Alternanthera pheloxirodes showed the beginning of a bloom in September and E. crassipes in October (Liu et al., 2004). Taihu Lake (China), is under threat from eutrophication due to the excessive amount of nutrients it receives from local industries and agricultural activities. However, Meiliang Bay is the major eutrophic area of this lake. The chemical oxygen demand was 4.63 mg/L in 1993. Total nitrogen and total phosphorus contents were 3.93 mg/L and 0.107 mg/L, respectively, in 1995. The Microcystis spp. among five major component phytoplankton species occupied 85% of the algal biomass and led to an algal bloom in summer that, in turn, affected the supply of water to the city of Wuxi (Weimin et al., 1997).

Compared to natural lakes, reservoirs tend to be more influenced by nutrients and other substances transported from the surrounding land. Lakes and reservoirs also differ in the amount of phytoplankton and aquatic plants (primary production) that can also be supported. In Developing Eutrophication Standards (DES) for Lakes and Reservoirs, the North American Lake Management Society (1992) states, "For the purposes of this document, perhaps the most important distinction between rivers, reservoirs and lakes is that of algal abundance per unit of phosphorus" (p. 9). Canfield and Bachman (1981) observed data from more than 700 natural lakes and reservoirs and compared their nutrient and response parameters. They found that reservoirs usually have substantially lower chlorophyll levels than natural lakes in the same phosphorus concentrations (Søballe and Kimmel, 1987). Cooke and Carlson (1989) reported mean chlorophyll-a values of 14.0 μg/L in natural lakes (n=309) and 10.0 μg/L in reservoirs (n=306). Based on these overall chlorophyll-a values, primary productivity appears to be lower in reservoirs than in natural lakes. Similarly, Søballe et al. (1992) found mean and median chlorophyll values for reservoirs in the southeastern U.S. to be significantly lower than those found in studies of natural lakes by Walker (1981) and Jones and Bachmann (1976). Lower productivity in reservoirs could occur because reservoirs tend to have higher concentrations of suspended solids and shorter hydraulic residence times compared to natural lakes (Søballe et al., 1992; Walker, 1984; 1985).

Many scientific experts have noticed that co-limitation of primary productivity by nitrogen and phosphorus is a common process in lakes and other water systems. As reported by Dodds et al. (1989), "statements that phosphorous is the major nutrient controlling primary productivity in freshwater systems should not be taken to mean that phosphorous is the only nutrient limiting productivity in all systems". The most commonly discussed of all nutrients, the three essential nutrients in fertilizer required for crop growth are nitrogen, phosphorus and potassium. These nutrients, when discharged into water bodies, promote phytoplankton (microscopic plants or algae) growth. Phytoplankton is primary producer, signifying the base of the food chain in all aquatic environments. The zooplankton feed upon phytoplankton, and small fish feed upon zooplankton. The smaller fish are consumed by large carnivorous fish. The growth of phytoplankton, or the primary productivity, is the first step in the food chain of a lake. The extent of algal production indicates to a certain degree the productive capacity of a lake. However, there are limits beyond which algal growth becomes detrimental to other aquatic life (Reutter, 1989).

Nutrients are necessary for all living cells; however, phosphorus is an important component of adenosine triphosphate, adenosine diphosphate, nicotinamide adenosine dinucleotide phosphate, nucleic acids, and phospholipids in cell membranes. Phosphorus may be stored in intracellular volutin granules as polyphosphates in both prokaryotes and eukaryotes. It is a limiting nutrient for algal growth in lakes and reservoirs. Phosphorus enters all water bodies continuously in runoff water and inlet streams. Phosphorus is also regularly lost from the water bodies through outlet streams and by assimilation into the sediments/mud. When a lake has anoxic bottom water in summer and stratifies, the top few millimeters of mud are chemically reduced to a condition that allows the phosphorus to be released back into the water. The bottom water thus becomes phosphorus rich. Water circulating around the lake due to winter storms brings the phosphorus-rich water to the surface, completing an annual cycle and fertilizing the lake for a spring plant bloom.

Phosphorus can be cycled some times through lakes and reservoirs, and transferred from one organism to another by food chains. Alternatively, it can sink to the bottom sediment as a part of fecal waste, a dead organism, or attached to a sinking particle. Once at the bottom of the lake or reservoir, phosphorus may become buried and unavailable to the system. Alternatively, rooted plants can transport phosphorus from the sediment into their tissues where, upon death, the phosphorus can be released back into the water (Horne and Goldman, 1994). Phosphorus in sediment may be released back into the system through chemical reactions; e.g., at pH values above 8, phosphate may disassociate from its particle and become soluble in water. Bottom-feeding fish and organisms that inhabit the bottom sediments such as worms and other aquatic organisms can also disturb the sediment, releasing phosphorus back into the water column. Phosphorus is released from lakes and reservoirs through the outflow to downstream waters (Hutchinson, 1957; Brönmark and Hansson, 2005).

Like phosphorus, Nitrogen (N) is also an essential nutrient for living organisms. It may come from natural sources, such as the decomposition of plants and animals, waste products from aquatic life within the water, urine and feces of wildlife in the catchments, or (in generally small amounts) mineral dissolution of rocks. Nitrogen also can enter lakes and reservoirs and is often of direct human origin (such as discharges from sewage treatment plants or leachate from septic systems) or is related to human activities (such as waste from poultry and livestock facilities, runoff of fertilizers, or nitrous oxides from fuel combustion). Nitrogen can be transported to lakes and reservoirs through atmospheric deposition (precipitation on the lake surface), runoff, or groundwater (Hutchinson, 1957; Wetzel, 2001).

Various chemical constituents of wild waters are thought to be an important factor in regulating the abundance, composition and geographical distribution of phytoplankton. Although phosphorus is mainly considered as the limiting nutrient for phytoplankton growth in every water system, the consequence of atmospheric nitrogen and its major role in the acidification of water can also be detrimental. Among nitrogen, phosphorus and silicon, nitrogen is usually considered as the primary limiting nutrient for the accumulation of phytoplankton diversity (Rabalais et al., 2002). Nitrogen is also an important component of chlorophyll, the green pigment that makes photosynthesis possible. It may limit phytoplankton production in temperate eutrophic waters, especially when phosphate concentrations are high (when nitrogen/phosphorus ratios are low). It is ever present in the water body in various forms. Phytoplanktons and other primary producers are able to utilize inorganic forms of nitrogen: nitrates (NO_3^-), nitrites (NO_2^-), ammonia (NH_3), and ammonium ions (NH_4^+) (Smith 1986). Some species of cyanobacteria (bluegreen algae) are also able to use nitrogen (N_2) directly from the atmosphere. Various forms of organic nitrogen (nitrogen that is bound to carbon-based molecules) may also become available in phytoplankton, like urea ($[NH_2]_2CO$), a soluble organic compound containing nitrogen that is excreted by urine and which can also be applied to the land as fertilizer, easily degrading into inorganic forms of nitrogen. Similarly, organic nitrogen found in plant and animal tissues can become available for use by primary producers if converted by bacteria into inorganic forms of nitrogen (Wetzel, 2001). Primarily, nitrogen can reduce from lakes and reservoirs through the outflow, in an exchange with groundwater, in the sediments and by denitrifying bacteria (e.g. converting NO_3^- to N_2) with subsequent loss of nitrogen gas (N_2) to the atmosphere (Hutchinson, 1957; Wetzel, 2001).

Including nutrients, other environmental factors also inhabited by plankton are heterogeneous, like temperature, irradiance and nutrient availability which are among the more obvious variables (Reynolds, 1984). The algal bloom caused by phosphorus input also modifies several abiotic factors in the water body. These factors directly rule the growth, diversity and density of the biotic components. The impact of algal bloom on any one or some of these factors indirectly influences the structure and characteristics of the water bodies. Ambient energy that can influence eutrophication dynamics may be fully investigated when long-term data with respect to time series are available (Lau and Lane, 2002).

Some microorganisms are also found to be proficient scavengers of phosphates from sewage sludge. A few strains of bacterium (Acinetobacter calcoaceticus) were found to remove substantial amounts of phosphates from an acetate medium-based pilot plant (Lawson and Tonhazy, 1980). Florentz and Hartemann (1984) also reported some varieties of A. calcoaceticus (var. lwoffi and var. anitratus), which is an efficient phosphate-removing bacteria in acetate-enriched pilot plants.

Eminent scientific groups have classified trophic status according to phosphorus absorption. Lakes with phosphorus concentrations below 0.01 mg/L are indicative of mesotrophic lakes, and phosphorus concentrations exceeding 0.02 mg/L are eutrophic lakes (Muller and Helsel, 1999). No rationale or state criteria have been established for concentrations of phosphorus compounds in water. However, to control eutrophication, the U.S. Environmental Protection Agency makes the following recommendations: Total phosphate (as phosphorus) should not exceed 0.05 mg/L in a stream at the point at which it enters a lake or reservoir and should not exceed 0.1 mg/L in streams that do not discharge directly into lakes or reservoirs (Muller and Helsel, 1999). Phosphate levels greater than 1.0 mg/L may interfere with coagulation in water treatment plants. As a result, organic particles that harbor microorganisms may not be completely removed before distribution. Eutrophic growth of aquatic plants seems to continue. Removal of detergent phosphates, by itself, will not usually accomplish such a large reduction because other inputs, such as runoff from agricultural lands, are much greater sources of phosphates. Those countries whose economies are based on agriculture have to take adequate measures in controlling the use of fertilizers, specifically phosphorus. Research and proper advice to farmers on the optimum requirement of the nutrients needs to be emphasized. Urban waste and sewage must be treated to reduce phosphorus before it is discharged into a water body.

Muller and Helsel (1999) suggested that, in order to reduce nutrient loads on lakes to within the limits permitted by the Organization for Economic Cooperation and Development, not only will all sewage inputs need to be prevented, and non-phosphate detergents used, but losses from agricultural land must be reduced.

The Three Gorges Project, the largest dam project in the world, is located in the mid-downstream area of the Yangtze River in Hubei and Sichuan Provinces, China. Despite the benefits of the dam in terms of power generation and flood control, the project has attracted attention for its potential impact on ecosystems and socio-economic stability. Especially in recent years, water bloom has occurred frequently in the backwater areas of the TGR in summer and autumn. Proper bio-indicators are expected to be provided to anticipate the trophic condition of the TGR, and to prevent the occurrence of water bloom based on previously effective measurements. The lack of functional bacteria data from the TGR is particularly worrying because environmental degradation in the TGR will probably be the most severe problem in the future.

Eutrophication of water components leads to harm for fisheries, recreation, industry, or for drinking, due to the adverse growth of algae and aquatic weeds and low oxygen level caused by decomposition. Also, many drinking water supplies throughout the world experience periodic huge surface blooms of cyanobacteria (Kotak et al., 1993). These blooms contribute to a wide range of water-related problems including summer fish kills, unpalatability of drinking water, and formation of trihalomethane during water chlorination (Kotak et al., 1994). Consumption of cyanobacterial blooms, or water-soluble neuro- and hepatotoxins (released when the blooms die), can kill life and may pose a serious health hazard to humans (Lawton and Codd, 1991). Mainly nitrogen (N) and carbon (C) are associated with accelerated eutrophication; special awareness has been paid to phosphorus (P), due to the difficulty in controlling the exchange of N and C between the atmosphere and a water body, and fixation of atmospheric N by some blue-green algae. Consequently, P is often the limiting element and its control is of the most importance in reducing the accelerated eutrophication of surface waters.

In practice, aquatic ecosystems are large sinks of nutrients such as phosphorus, nitrogen and carbon that are stored in the sediments of the water body, which is very important to water quality management. However, sinks of nutrients in aquatic or wetland systems also play a major role in regulating the transformation of nitrogen and carbon (Moore et al., 2003) as well as other biologically regulated elements such as sulphur. With the liberation of nutrients from the lake, reservoir or wetland systems, the subsequent inundation is controlled by leaching from the sediments and the decomposition of organic matter such as vegetation and detritus. Water temperature, inundation frequency, timing, duration and areal extent (St. Louis et al., 2003) have also been acknowledged to impact the above systems and affect the release, mineralization and transport of nutrients from aquatic sediments to surface waters as well as the potential transfer to hydrologically based aquatic ecosystems.

Aquatic ecosystems are sensitive to amplification of the external burden of phosphorus (organic and inorganic) from both anthropogenic and natural sources. Organic phosphorus is always associated with lake sediments and vegetation (Hogan et al., 2004). However, inorganic phosphorus is adsorbed in particles of soil or bound with lake sediments, held in complexes with iron (Fe^{3+}), aluminum (Al^{3+}) and calcium (Ca^{2+}) (Song et al., 2007). Changes in hydrological features may affect the water system. The adsorption ability of sediments decreases due to alteration from inundation. However, the hydrolysis of Fe and Al phosphate complexes gradually increases leading to the release of phosphorus into the overlying water column through leaching. As water levels recede, the subsequent decomposition of organic matter is accelerated as oxygen is initiated and phosphatase enzyme activity gradually increases, which regulates the mineralization of organic phosphorus (Song et al., 2007). On the other hand, naturally extracted inorganic phosphorus from inundated water sediments is liberally transported to adjacent water bodies by surface runoff during precipitation events (Young and Ross, 2001; Kieckbusch and Shrautzer, 2007).

Reducing P inputs into lakes may not always attain the preferred or even predictable water quality improvement, due to the sustained input of P from other sources. The direct input of P in rainfall into lakes may be sufficient to enhance the algal growth in certain states. Elder (1975) projected that rainfall P may account for up to 50% of P entering Lake Superior, and the enrichment of lakes in Ontario (Schindler and Nighswander, 1970) and Wisconsin (Lee, 1973) has been attributed to rainfall P. In addition, the release of P from sediment can sustain the growth of aquatic biota for several years after its deposition (Ahlgren, 1977; Jacoby et al., 1982). Therefore, the same form of in-lake management to reduce aquatic bioproductivity may be necessary and cost-effective. Obviously, lake use has an impact on desirable water quality goals, which will require contradictory management. Watershed management often becomes more complex with multiple-use lakes and streams, which tend to dominate U.S. waters. For example, a reservoir may have been built for water supply, hydropower, and/or flood control and, although not a primary purpose, recreation is often considered a benefit, with aesthetic enhancement (including property value) as an additional fringe benefit. Nutrients might be lost dissolved in water and involved with eroded soil/sediment in surface runoff and dissolved in leaching water. Losses of the major nutrients nitrogen (N) and phosphorus (P) from terrestrial to water resources cause water quality concerns with regard to the health of humans and aquatic systems as well as damaging the utility of water resources.

In aquatic ecosystems, the primary producer is an imperative component enhancing microbial activity, where nitrate (NO_3) and nitrite (NO_2) act as electron acceptors under anaerobic conditions during the decomposition of organic matter (D'Angelo and Reddy 1999). Generally, nitrogen is continually cycled and transformed by microbial activity (Mitsch and Gosselink 1993). Nitrogen can exist as ammonium (NH_4^+), NO_3, NO_2 or be bound in organic forms; under anoxic conditions, the partial or complete denitrification of inorganic nitrogen by microbial activity may be released as nitrous oxide (N_2O) or nitrogen gas (N_2) (Anderson, 2004). While organic matter decomposes, organic nitrogen is readily mineralized into NH_4^+ and it may dispersed into the water column as aquatic soils are inundated. If mineralization rates of organic nitrogen exceed denitrification rates in water sediments or macrophyte nitrogen-limitation becomes saturated, the transport of inorganic nitrogen to hydrologically connected aquatic systems may occur (Hemond, 1983).

The retention potential of wetland systems is quite significant, as a large portion of nitrogen entering aquatic ecosystems through precipitation and surface runoff is lost to the atmosphere (Gergel et al., 2005) or assimilated by wetland macrophytes (Laiho and Vasander, 2003). Despite this, the eutrophication of aquatic ecosystems by nitrogen loading, specifically inorganic forms such as NO_3, is readily documented (Camargo and Alonso, 2006). While most temperate freshwater systems are phosphorus limited, increases in total nitrogen concentrations may shift the phytoplankton community structure. Increases in cyanobacterial biomass have been correlated with both total nitrogen and total phosphorus concentrations, an important implication for drinking water reservoirs.

Organic carbon is also an important component of the microbial food web, providing an energy source to power many of the biologically-mediated nutrient cycling processes. While sediment of the water systems generally acts as a sink of organic carbon, the inundation of soils can lead to increased leaching of organic carbon (Asada et al., 2005) and the subsequent transport to the overlying water column (Moore et al., 2003; Corstanje and Reddy, 2004; St. Louis et al., 2004). The inundation of sediment also stimulates the mineralization of organic carbon and the subsequent release of carbon dioxide (CO_2) (Kelly et al., 1997; St. Louis et al., 2003). Under very reducing environments, the CO_2 is also subject to methanogenesis, being reduced to methane (CH_4) (Mitsch and Gosselink, 1993; Kelly et al., 1997; St. Louis et al., 2003). In addition to increasing the export of organic carbon to hydrologically connected aquatic ecosystems (Moore et al., 2003; St. Louis et al., 1996), both the methanogenesis and mineralization processes can switch a wetland system from being a carbon sink to a carbon source, significantly increasing the emission of CO_2 and CH_4 after inundation (Kelly et al., 1997).

In wetland systems, dissolved organic carbon is often transported to adjacent and hydrologically connected aquatic ecosystems, carrying with it heavy metals (St. Louis et al., 2004). As wetland soils are inundated and organic carbon is processed, decomposition, mineralization, respiration and hydrologic transport all influence the potential export of carbon to hydrologically-connected aquatic ecosystems.

The management of phosphorus, nitrogen and organic carbon in aquatic ecosystems is an important factor in controlling cultural eutrophication (Schindler, 1977), especially in drinking water reservoirs, where increases in total phosphorus and nitrogen concentration can significantly increase algal biomass, potentially creating blooms of taste and odour causing species or cyanobacteria. Total phosphorus concentration is also a predictor of other odour-producing compounds in drinking water reservoirs. However, increased concentrations of organic carbon react with bromine and chlorine to produce carcinogenic disinfection byproducts and may increase bacterial biomass within distribution systems (LeChevallier et al., 1996).

1.2 Water Quality and Eutrophication

The term eutrophication refers to the natural and artificial addition of nutrients to water bodies and to the effect that these added nutrients have on water quality (Vollenweider and kerekees, 1980). The characteristic of eutrophication is the excessive growth of algae, although many components and elements contribute to the algal growth. The most crucial nutrients are usually required in a ratio of 106 units of carbon to 10 units of nitrogen and 1 unit of phosphorus (Ryther and William, 1971). It has been shown by Liebig's Law of Minimum that the component in shortest supply will control the growth rate (Shu, 1982; Hecky, 1988). Commonly, nitrogen and phosphorous are both considered as the most frequent limiting factors in the aquatic environment, and investigations of the nutrient profile of the water body and also algae have afforded threshold estimates of nitrogen and phosphorus concentration that limit algal growth. When the total nitrogen (TN) to total phosphorus (TP) ratio in water exceeds 12 during the spring period, then phosphorus should be the limiting nutrient (Dillon and Rigler, 1974). The primary symptom of eutrophication is a high concentration of TNand of TP in the water. Sawyer (1947) suggests that TP concentration in excess of 0.01 mg/L and TN concentration above 0.3 mg/L are adequate to cause nuisance algal blooms.Since eutrophication is increased nutrient input, any activity in the watershed of a lake that increases nutrient input causes eutrophication. New Hampshire revealed that phosphorus export from agricultural lands is 5 times more than from forested lands, and from urban areas it may be more than 10 times greater. Storm water runoff from the developed land areas is the major source of nutrients for most lakes and resorvoirs. Other activities that contribute to eutrophication are lawn and garden fertilizers, faulty septic systems, washing with soap in or near the lake, erosion into the lake, dumping or burning leaves in or near a lake, and feeding ducks.

A eutrophic lake is typically shallow with a soft and muddy bottom. Rooted plant growth is abundant along the lake and algal blooms are not unusual. Water clarity is not good and the water often has a tea color. A shallow, well-mixed water column usually prevents stratification, so bottom waters are usually oxic in condition, and frequent sediment resuspension inhibits photosynthesis below 0.5–1 m water depth (Dickman et al., 1998). Cyanobacteria populations (Microcystis and Anabaena) have increased and can comprise up to 85% of summer phytoplankton biomass (Chen et al., 2003). Many natural water bodies are described as oligotrophic, for they have clear water ecosystems in which primary and secondary productivity is limited by a shortage of major nutrients (Beeby, 1995). These oligotrophic water bodies, if brought under natural succession, need thousands of years to become eutrophic. The enrichment of aquatic ecosystems through the discharge of human wastes from settlements and excessive fertilizers from agricultural lands brings down the water bodies in an undesirably increased rate of eutrophication.

Among various natural factors, water is an essential, life-supporting factor in every living cell, individual organism, ecosystem and universe. Water quality is deteriorating gradually due to the rapid progress of scientific organizations. Unfavorable fluctuations in the physicochemical characteristics of the water body bring water contaminants, which affect the planktonic flora of the water body. Planktonic flora including micro organisms are inadequately supplied by the available phosphorus or nitrogen. According to Likens et al. (1977), oligotrophic water bodies contain less than 5–10 μg/L of phosphorus and less than 250–600 μg/L of nitrogen. The mean primary productivity in oligotrophic water has been noticed between 50–300 mg carbon/(m^2·day). In moderately eutrophic water bodies, the phosphorus content is 10–30 μg/L and the nitrogen content is 500–1100 μg/L. The primary productivity in eutrophic water is reported to be above 1 g carbon/(m^2·day). If excessive amounts of phosphorus and nitrogen are added to the water, aquatic organisms and plants can grow over a large area and eutrophication occurs. When an organism dies, it is naturally decomposed by bacteria and the decomposers use up the dissolved oxygen in the water body. The dissolved oxygen concentrations often drop too low for aquatic life to breathe, leading to life kills (Murphy, 2002). It is well known that eutrophication is a natural process in lakes as well as reservoirs. Nitrification is usually included in water quality management due to its significance in mediating the contents of dissolved oxygen, ammonia, nitrate-nitrogen and phosphates (Deb and Bowers, 1983).

Nitrification, microbial-mediated oxidation of ammonia to nitrate, is a key process in the N cycling in every water component. It is a two-step process; firstly an ammonia-oxidizing stage that transforms NH_4^+ to NO_2^- by ammonia-oxidizing bacteria (AOB); secondly the nitrite-oxidizing stage that transforms NO_2^- to NO_3^- by nitrite-oxidizing bacteria (NOB), in which the ammonia-oxidizing stage is the critical stage controlling the nitrification (Costa et al., 2006). Hyper-nitrification caused by the heavy loads of N and P, is greatly assisted by the N and P cycling in the natural environment (Jergensen and Richardson, 1996; Paerl, 1997). It is often assumed that nitrification occurs in the water column and that the process follows first-order kinetics with rates calculated as a function of water column ammonia contents (Ambrose et al., 1993; Scott and Abumoghli, 1995). However, nitrification in freshwater is known to take place mainly on suspended particles (SPs) and bed sediments (Bonnet et al., 1997). Consequently, the investigation of the relationship between SPs and the nitrification rate in the aquatic environment is of importance for developing the ecosystem model of nitrogen biogeochemical cycling (Kittiwanich et al., 2007) and water quality management.

Even though the eutrophication process in lakes has not yet led to a significant deterioration in the water quality, the ecological balance within lakes is extremely fragile owing to the high nutrient loadings. Therefore, lake water management must focus on reducing the nutrient input into the lakes from the streams that drain the vast rural areas of Jiangsu and Zhejiang provinces and the Great Shanghai Metropolitan Area.

1.3 Organization of the Book

This book mainly focuses on the distribution and transformation law of nutrients in large-scale lakes and reservoirs. The readers will include environmental engineers, environmental scientists and ecological chemistry researchers.

- This book firstly introduces the distribution and transformation of nutrients (Chapters 1 and 2);
- Hydrodynamic effects, biological effects and chemical effects on the distribution and transformation of nutrients are described in Chapters 3, 4, and 5, respectively;
- Mathematical modeling and numerical simulation are developed to predict water quality changes in Chapter 6;
- The eutrophication risk assessment for large-scale lake and reservoirs is discussed in Chapter 7.

References

Ahlgren I (1977) Role of sediments in the process of recovery of a eutrophicated lake. In: Golterman H. (ed.), Interaction between Sediment and Freshwater. Proceedings of an International Symposium Held at Amsterdam, The Netherlands, September 6-10, 1976. Dr. W. Junk B. V. Publishers, The Hague.

Anderson HE (2004) Hydrology and nitrogen balance of a seasonally inundated Danish floodplane wetland. Hydrol Process 18:415-434.

Asada TB, Warner G, Schi SL (2005) Effects of shallow flooding on vegetation and carbon pools in boreal peatlands. Appl Veg Sci 8:199-208.

Beeby A (1995) Applying Ecology. Chapman & Hall, London.

Bonnet C, Volat B, Bardin R, Degrange V, Montuelle B (1997) Use of immunofluorescence technique for studying a nitrobacterial population from wastewater treatment plant following discharge in river sediments: first experimental data. Water Res 31:661-664.

Brönmark C, Hansson LA (2005) The Biology of Lakes and Ponds, 2nd Edition. Oxford University Press Inc., New York, NY, p. 285.

Camargo JA, Alonso A (2006) Ecological and toxicological effects of inorganic nitrogen pollution in aquatic ecosystems: A global assessment. Environ Int 32:831-849.

Canfield D, Bachmann R (1981) Prediction of total phosphorus concentrations, chlorophyll a, and Secchi depths in natural and artificial lakes. Can J Fish Aquat Sci 38:414-423.

Chen W, Chen Y, Gao X, Yoshida I (1997) Eutrophication of Lake Taihu and its control. Int Agric Eng J 6:109-120.

Chen Y, Qin B, Teubner K, Dokulil M (2003) Long-term dynamics of phytoplankton assemblages: Microcystis-Domination in Lake Taihu, a large shallow lake in China. J Plankton Res 25:445-453.

Cooke GD, Carlson RE (1989) Reservoir Management for Water Quality and THM Precursor Control. AWWA Research Foundation and American Water Works Association, Denver, CO, p. 387.

Corstanje R, Reddy KR (2004) Response of biogeochemical indicators to a drawdown and subsequent reflood. J Environ Qual 33:2357-2366.

Costa E, Perez J, Kreft JU (2006) Why is metabolic labor divided in nitrification? Trends Microbiol 14:213-219.

D'Angelo EM, Reddy KR (1999) Regulators of heterotrophic microbial potentials in wetland soils. Soil Biol Biochem 31:815-830.

Deb A, Bowers D (1983) Diurnal water quality modeling: a case study. J Water Pollut Control Fed 55:1473-1488.

Dickman M, Pu P, Zheng C (1998) Taihu Lake: past, present and future. J Lake Sci (China) 10(Suppl):75-83.

Dillon PJ, Rigler FH (1974) The phosphorus-chlorophyll relationship in lakes. Limnol Oceanogr 19:767-773.

Dodds WK, Johnson KR, Priscu JC (1989) Simultaneous nitrogen and phosphorous deficiency in natural phytoplankton assemblages: theory, empirical evidence, and implications for lake management. Lake Reserv Manag 5:21-26.

Elder FC (1975) International Joint Commission Program for Atmospheric Loading of the Upper Great Lakes. Second Interagency Committee on Marine Science and Engineering Conference on the Great Lakes, Argonne, Illinois.

Florentz M, Hartemann P (1984) Screening for phosphate accumulating bacteria isolated from activated sludge. Environ Technol Lett 5:457-463.

Ford DE (1990) Reservoir transport processes. In: K.W. Thornton, B.L. Kimmel, F.E. Payne (eds.), Reservoir Limnology: Ecological Perspectives. John Wiley & Sons, Inc., New York, NY, pp. 15-41.

Gergel SE, Carpenter SR, Stanley EH (2005) Do dams and levees impact nitrogen cycling? Simulating the effects of flood alterations on foodplain denitrification. Glob Change Biol 11:1352-1367.

He F, Wu ZB, Qiu DR (2002) Allelopathic effects between aquatic plant (Potamogeton crispus) and algae (Scenedesmus obliquus) in enclosures at Donghu Lake. Acta Hydrobiologica Sinica 26:421-424.

Hecky RE (1988) Nutrient limitation of phytoplankton in freshwater and marine environments: a review of recent evidence on the effects of enrichment. Limnol Oceanogr 3(4, part2):792-822.

Hemond HF (1983) The nitrogen budget of Thoreau's Bog. Ecology 64(1):99-109

Hogan DM, Jordan TE, Walbridge MR (2004) Phosphorus retention and soil organic carbon in restored and natural freshwater wetlands. Wetlands 24:573-585.

Horne AJ, Goldman CR (1994) Limnology. McGraw-Hill, Inc., New York, NY.

Hutchinson GE (1957) A Treatise on Limnology Vol. 1 Geography, Physics and Chemistry. John Wiley & Sons, New York, NY.

Jacoby JM, Lynch DD, Welch EB, Perkins MS (1982) Internal phosphorus loading in a shallow eutrophic lake. Water Res 16:911-919

Kelly CA, Rudd JWM, Bodaly RA, Roulet NP, St. Louis VL, Heyes A, Moore TR, Schi S, Aravena R, Scott KJ, Dyck B, Harris R, Warner B, Edwards G (1997) Increases in fluxes of greenhouse gases and methyl mercury following flooding of an experimental reservoir. Environ Sci Technol 31:1334-1344.

Kittiwanich J, Yamamoto T, Kawaguchi O, Hashimoto T (2007) Analyses of phosphorus and nitrogen cyclings in the estuarine ecosystem of Hiroshima Bay by a pelagic and benthic coupled model. Estuary Coastal Shelf Sci 75:189-204.

Kotak BG, Kenefick SL, et al. (1993). Occurrence and toxicological evaluation of cyanobacterial toxins in Alberta lakes and farm dugouts. Water Res 27(3): 495-506.

Kotak BG, Prepas EE, Hrudey SE (1994) Blue-green algal toxins in drinking water supplies research in Alberta. LakeLine 14(1):37-40.

Laiho R, Vasander H (2003) Dynamics of plant-mediated organic matter and nutrient cycling following water-level drawdown in boreal peatlands. Glob Bio Geochem Cycle 17(2):1-11.

Lau SSS, Lane SN (2002) Biological and chemical factors influencing shallow lake eutrophication: a long-term study. Sci Total Environ 3:167-181.

Lawson EN, Tonhazy NE (1980) Changes in morphology and phosphate uptake patterns of Acinetobacter calcoaceticus strains. Water SA 6:105-112.

Lawton LA, Codd GA (1991) Cyanobacterial (Blue-Green Algal) toxins and their significance in UK and European waters. Water Environ J 5(4): 460-465.

LeChevallier, Welch MWNJ, Smith DB (1996) Full-scale studies of factors related to coliform regrowth in drinking water. Appl Environ Microbiol 62(7):2201-2211.

Lee GF (1973) Role of phosphorus in eutrophication and diffuse source control. Water Res 7:111-128.

Likens GE, Bormann FH, Pierce RS, Eaton JS, Johnson NM (1977) Biogeochemistry of a Forested Ecosystem. Springer-Verlag, New York.

Liu C, Wu G, Yu D, Wang D, Xia S (2004) Seasonal changes in height, biomass and biomass allocation of two exotic aquatic plants in a shallow eutrophic lake. J Freshw Ecol 19:41-45.

Mitsch WJ, Gosselink JG (1993)Wetlands, 2nd Ed. Van Nostrand Reinhold, New York.

Moore TR, Matos L, Roulet NT (2003) Dynamics and chemistry of dissolved organic carbon in Precambrian Shield catchments and an impounded wetland. Can J Fish Aquatic Sci 60:612-623.

Muller DK, Helsel DR (1999) Nutrients in the nation's water—Too much of good thing? Circular 1136. U.S. Geological Survey, Denver.

Murphy S (2002) General information on phosphorus. City of Boulder/USGS Water Quality Monitoring.

Paerl HW (1997) Coastal eutrophication and harmful algal blooms: importance of atmospheric deposition and groundwater as 'new' nitrogen and other nutrient sources. Limnol Oceanogr 42:1154-1165.

Rabalais NN, Galloway J, Cowling E (2002) Nitrogen in aquatic ecosystems. 2nd International Nitrogen Conference on Optimizing Nitrogen Management in

Food and Energy Productions, and Environmental Change, Potomac, Maryland, USA, October 2001. Ambio 31:102-112.

Reutter JM (1989) Lake Erie: Phosphorus and Eutrophication. Fact Sheet 015. Ohio Sea Grant College Program, Columbus.

Reynolds CS (1984) The Ecology of Freshwater Phytoplankton. Cambridge Univ. Press, Cambridge.

Ryther JH, William MD (1971) Nitrogen, phosphorus and eutrophication in a coastal marine environment. Science 171:1008-1013.

Sawyer CH (1947) Fertilization of lakes by agricultural and urban drainage. J. of New England Water Works Ass. 61:109-127.

Schindler DW (1977) Evolution of phosphorus limitation in lakes. Science 195:260-262.

Schindler DW, Nighswander JE (1970) Nutrient supply and primary production in Clear Lake, eastern Ontario. J Fish Res Board Can 27:260-262.

Scott JA, Abumoghli I (1995) Modelling nitrification in the river Zarka of Jordan. Water Res 29:1121-1127.

Shu JH (1982) Lake eutrophication and its control. J. Environ. Quality 10:23-32.

Smith VH (1986) Light and nutrient effects on relative biomass of blue green algae in lake phytoplankton. Can J Fish Aquatic Sci 43:148-153.

Søballe DM, Kimmel BL (1987) A large scale comparison of factors influencing phytoplankton abundance in lakes, rivers and impoundments. Ecol 68:1943-1954.

Søballe DM, Kimmel BL, Kennedy RH, Gaugush RF (1992) Biodiversity of the southeastern United States: Aquatic communities. Res 421-474.

Song H, Heinz U (2008) Suppression of elliptic flow in a minimally viscous quark–gluon plasma. Phys Lett B 658(5):279-283.

St. Louis VL, Rudd JWM, Kelly CA, Beaty KG, Flett RJ, Roulet NT (1996) Production and loss of total mercury from boreal forest catchments containing different types of wetlands. Environmental Science and Technology 30:2719-2729

St. Louis VL, Partridge AD, Kelly CA, Rudd JWM (2003) Mineralization rates of peat from eroding peat islands in reservoirs. Biogochemistry 64:97-109.

St. Louis VL, Rudd JWM, Kelly CA, Bodaly RA, Paterson MJ, Beaty KG, Hesslein RH, Heyes A, Majewski AR (2004) The rise and fall of mercury methylation in an experimental reservoir. Environ Sci Technol 38:1348-1358.

Vollenweider RA, Kerekes J (1980) The loading concept as a basis for controlling eutrophication: Philosophy and preliminary results of the OECD program on eutrophication. Progr. Water Technol 12:5-38.

Walker WW, Jr (1981) Empirical Methods for Predicting Eutrophication in Impoundments, Report 1, Phase I. Data Base Development. Tech. Rep. E-81-0. EWQOS, U.S. Army Corps of Engineers, Vicksburg, MS.

Walker WW, Jr (1984) Empirical predication of chlorophyll in reservoirs. U.S. Environmental Protection Agency. Lake Res Manag 292-297.

Walker WW, Jr (1985) Empirical Methods for Predicting Eutrophication in Impoundments, Rep. 3, Phase II. Model Refinements. Tech. Rep. E-81-9. U.S. Army Engineer Waterways Experiment Station, Vicksburg, MS.

Wetzel RG (2001) Limnology (3rd ed.). Academic Press, New York, p. 1006.
Wisconsin Dept. Natural Resources (1986) Nonpoint source pollution abatement program. Wisconsin Administrative Code NR 120. Madison, Wisconsin
Young EO, Ross DS (2001) Phosphate release from seasonally flooded soils: a laboratory microcosm study. J Environ Qual 30:91-101

2

Eutrophication and Distribution of Nitrogen and Phosphorus

2.1 Overview

Eutrophication is the phenomenon of phytoplankton bloom caused by excessive nutrients (N, P and other inorganic salt) in water environment. The most widely accepted definition of eutrophication is that which describes the nutrient enrichment of waters which results in the stimulation of an array of symptomatic changes, among which increased production of macrophytes and algae, deterioration of water quality and other symptomatic changes are found to be undesirable and interfere with water uses. Some of problems were brought about by eutrophication including turbid waters, bad smell, anoxic conditions and chironomid and Culex midge plagues, even more intoxicating phenomena for animals and people (Beat et al., 2008; Cecilia et al., 2005; Datta et al., 1999).

The study of eutrophication has tended to focus on reservoirs and lakes, and recognition is growing of the problems of eutrophication leading to algal blooms. Lots of factors have been demonstrated to affect algal growth in reservoirs and lakes, the most important of which are sunlight (Swale, 1969), flow (Lack, 1971), temperature (Hasnain et al., 1990), turbidity and nutrients (Lund, 1970). Attention has been focused on the role of nutrients in controlling the algal growth, as they are considered the most difficult to control (Schindler, 1978). Such environmental eutrophication problems are generally considered to be the consequence of enhanced nutrient loadings (Helen et al., 2006; Kristina et al., 2005; Young et al., 1999; Gao et al., 2008).

In the aquatic ecosystem, it is generally believed that nutrients with restrictive effects exist, which means that organisms would compete with each other, when the content of the nutrient was reduced. The studies on eutrophication by the International Economic Cooperation and Development Organization (OECD) indicated that nitrogen and phosphorus can be made to be the important nutrients (Carpenter et al., 1998). According to the chemical composition analysis of algae, the molecular formula of algae is $C_{106}H_{236}O_{110}N_{16}P$. As the Liberg Minimum Value Law shows, the growth of a plant would be decided by the nutrient.

Therefore, nitrogen and phosphorus are the two elements with the smallest percentage by weight in the molecular formula of algae. In the eutrophication of lakes and reservoirs, the amount of eutrophication restricted by the concentration of phosphorus is 80%, and 10% is related to the concentrations of nitrogen and phosphorus. For the other 10% in the lakes and reservoirs, eutrophication may be due to other environmental factors. All of this indicates that the nitrogen and phosphorous in the water environment are the primary limiting nutrients.

Due to rapid development, great harm, difficult control and slow restoration, eutrophication has gradually become a global water pollution problem (Hans, 2006). And because the nitrogen and phosphorous are the primary limiting nutrients, the study of the distribution of nitrogen and phosphorous in lakes and reservoirs is becoming a world concern, with eutrophication problems evident in many countries, such as Belgium, Denmark, Germany, Italy, The Netherlands and so on. In the Gulf of Finland, the northern Baltic Sea, the range of concentrations of nitrogen and phosphorous were 0.25–1.2 mg/L and 0.014–0.44 mg/L in the water (Cecilia et al., 2005), respectively. In Lagunitas Creek and Tomasini Point, the contents of TN and TP were 231.6 and 15.29 mmol/kg in the sediments (Vink et al., 1997), and most of the phosphorus (70%–40%) in the sediments was found in the organic and residual speciation. In two anthropogenically influenced estuarine systems in southwest England, the concentrations of DON in both estuaries were generally lower than 80 μM (El-Sayed et al., 2008). DON contributed 38%±22% (range 4%–79%, Yealm) and 36%±17% (range 4%–84%, Plym) to the TDN. And relative to winter and spring, the DON was a larger fraction of the TDN during the summer and autumn.

As known, water environment of the Three Gorges Reservoir is important in China. The Three Gorges Reservoir is located on the Yangtze River which is the largest river in China. The water resources of the Three Gorges Reservoir area are rich, and there are many secondary rivers. With the rapid development of the urban economy, lots of pollutants were discharged into water and caused pollution of the secondary rivers, which caused the water quality to decrease. After the impoundment of the Three Gorges Reservoir, the conditions in the reservoir changed. The flow rate of tributaries in backwater areas decreased because of the jacking effect of the high water level in the main stream. The ability of the rivers to transport N and P was reduced; the nutrients may have accumulated there. As a result, the study of the distribution of N and P in the Three Gorges reservoir is necessary for the prevention of eutrophication there.

2.2 Eutrophication

Eutrophication is a phenomenon where phytoplankton and other aquatic plants bloom; the transparency and dissolved oxygen in the water decrease, and the condition of the water changes from a state of poor nutrition to a state of high nutrition, which is due to excessive nutrients (N, P and other inorganic salt) discharging into reservoirs, lakes, estuaries and gulf waters, which are relatively closed and slow-moving waters.

Eutrophication includes natural and artificial eutrophication. The occurrence of natural eutrophication is due to lakes, reservoirs and other waters receiving the nutrients from natural rainfall and leaching of the soil, which lead to the increase of nutrients in water. And this is also due to the rapid development of industrial and agricultural production, the emissions of nitrogen and phosphorus discharged directly into the water environment. All of these, which increase the amount of nutrients, cause artificial eutrophication. The reason for natural and artificial eutrophication is the enrichment of nutrients such as nitrogen and phosphorus in the water environment. This leads to algae and other phytoplankton reproduction booming, and dissolved oxygen dropping, which eventually causes a large number of the fish or other aquatic organisms to die and water quality to deteriorate.

The main appearance of eutrophication in water is seen in the blooming of plankton. Some of the manifest problems brought about by plankton algal biomass include turbid waters, anoxic conditions, bad smell and chironomid and Culex midge plagues (Beat et al., 2008; Cecilia et al., 2005; Datta et al., 1999). When algal growth increases, the content of dissolved oxygen will decrease, with an increased incidence of toxic cyanobacteria blooms and a decrease in the abundance of species. Such eutrophication problems are generally considered to be the consequence of enhanced nutrient loadings (Helen et al., 2006; Kristina et al., 2005; Young et al., 1999; Gao et al., 2008).

Due to rapid development, great harm, difficult control and slow restoration, eutrophication has gradually become a global water pollution problem (Hans, 2006). Nowadays, significant signs of eutrophication have been reported globally, with major affected areas including the Baltic, Adriatic and Black Seas, and the coastal waters of Japan, Australia, USA and China. According to the survey, many reservoirs and lakes in China may be eutrophicated, which have affected urban water supply security and sustainable development of the local economy. The distribution of nutrients in Taihu Lake, which is an important drinking water source for Wuxi City and a famous scenic spot in China, was studied by lots of researchers. The concentration ranges of DIN, NO_3^--N and NH_4^+-N were 0.2–5.1, 0.1–2.9 and 0.07–0.99 mg/L, respectively, in 2003 (Yang et al., 2007). The Daliao River system consists of the Daliao River, Hun River and Taizi River. The concentration of P in the sediments of the Daliao River ranged from 230 to 841 mg/kg in 2007 (Wang P et al., 2009). In Poyang Lake in China, the content of total phosphorus in the surface sediments ranged from 688.29 to 825.36 mg/kg (Xiang and Zhou, 2011), and inorganic phosphorous was the major phosphorus speciation.

The Three Gorges Dam, in China, is the world's largest hydroelectric project, measuring 2335-m long and 185-m high, and the reservoir created by it had an area of 1080 km^2 in 2009. The Three Gorges Dam is 27 miles upstream from Yichang City, at Sandouping Town, 38 km upstream from the Gezhouba Dam Lock, and inside the third of the Three Gorges. The construction of the Three Gorges Dam began in 1993. Now the Three Gorge Reservoir is completed, it is capable of containing 39.3 billion m^3 of water, which has lifted the water surface 175 m above sea level. The dam will cause changes in the discharging pattern of the Yangtze River. The flow rate downstream of the dam decreases in October because of impoundment and increases from January to May. The rest of the time the monthly flow remains relatively unchanged. The construction of the Three Gorges Reservoir will have a profound impact on the environment, wherein eutrophication is a potential threat. Phosphorus and nitrogen are the major nutrients for aquatic ecology, and their excess supply can lead to eutrophication. When external loading of P and N increase, the sediments as a pool can adsorb it. However, after external loading is reduced, the sediments as a source will release adsorbed P and N into water. Therefore, intensive study of P and N in water and sediments would help to predict and further prevent the occurrence of eutrophication.

According to the research reported by Li et al. (2005), after the Three Gorges Reservoir was finished, the nitrogen and phosphorus levels reached the level where eutrophication occurred in main stream of the Three Gorges reservoir area. But due to the fact that the depth of most of the river was an average of 70 m, and light and temperature levels were not high enough at the bottom, the condition there was not suitable for the growth of algae, especially in the areas in front of the dam, where the flow rate was lower than 0.2 m/s, and the depth was greater than 100 m. Therefore, in the main stream of the Three Gorges Reservoir area eutrophication occurred with difficulty.

The conditions in the secondary rivers in the Three Gorges Reservoir area are different. After the Three Gorges Reservoir was finished, the depth of water at the end of the river backwater areas was generally 10 m. Moreover, the flow rate here underwent great changes, when the flow rate dropped to a similar rate to natural lakes. The conditions for eutrophication were the same as in reservoirs or lakes. Therefore, when the nitrogen and phosphorus concentrations in the secondary rivers reached or exceeded the level for eutrophication, this became a potential risk (Beat et al., 2008). The rivers like the Kaixian, Daning, Hong and other waters like the typical backwater areas of the secondary rivers, are the most sensitive areas for eutrophication in the Three Gorges reservoir.

The Longchuanjiang River, which originates in Nanhua County in southwest China, is located in the upper stream of the Yangtze River. The concentrations of NO_3^-, NH_4^+, DN and TN there were 1037, 835, 1672 and 1935 μg/L. And the concentrations of DP, PAP and TP in the water were 33, 609 and 642 μg/L (Lu et al., 2011).

At about 400 km from the mouth of the Yangtze River, Datong is the gauging station for the Yangtze River's total water and sediment discharge. Hydrographic and nutrient data has been available for a few decades. In the water at Datong station (Duan et al., 2008), NO_3^-, accounting for 48%–79% of total N, was the main component of N for all the seasons. Conversely, NH_4^+ and NO_2^- were minor forms of TN during the flood and pre-flood periods. The NO_3^- concentrations ranged from 70 to 97 mmol/L during the summer to pre-flood period, and 1.8–22.1 μmol/L and 0.2–2.5 μmol/L for the NH_4^+ and NO_2^- concentrations. The DIP concentrations ranged from 0.2 mmol/L during the summer flood to 1.0 mmol/L during the winter low-flow period.

As known, the Yangtze estuary has received high loading of anthropogenic nutrients from the Yangtze River. The MBP concentrations (the content of phosphorus released from sediment during acidic digestion) of the Yangtze estuarine sediment samples, with an average concentration of 17.14 ng/kg, ranged from 1.93 to 94.86 ng/kg (Feng et al., 2008). The concentrations of TP, IP and OP in the Yangtze estuary sediment samples were also analyzed. TP concentrations were\ high, ranging from 267.8–525.2 mg/kg. OP and IP concentrations ranged from 20.7 to 173.5 and from 224.0 to 408.1 mg/kg, respectively. And TN concentrations ranged from 0 to 0.81 g/kg.

As known, there are lots of tributaries of the Yangtze River. The nutrients and other pollutants from tributaries flow into the Yangtze. The Minjiang River and Tuojiang River are two important tributaries of the Yangtze. As shown in Table 2.1, the concentrations of DP were 0.048–0.063, 0.057 and 0.262 mg/L in the water from the Yangtze, Mingjiang and Tuojiang, respectively (Cao et al., 2011). And in sediments, the concentrations of TP, AP and FP were 890–954, 7.1–10.2 and 83–110 mg/kg in the Yangtze, 1010, 35.7 and 76 mg/kg in the Mingjiang, and 1130, 82.8 and 85 mg/kg in the Tuojiang. The highest DP concentration occurred in the Tuojiang, and the lowest was for the Yangtze. Obviously, the Tuojiang has been severely polluted because it drains urban areas (Cao et al., 2011).

Table 2.1 The concentrations of dissolved phosphorus (DP) in the water samples and total phosphorus (TP), available phosphorus (AP), phosphorus-fixing capacity (FP) in the sediment samples from the Yangtze River and tributaries (Cao et al., 2011)

River	Water		Sediments	
	DP (mg/L)	TP (mg/kg)	AP (mg/kg)	FP (mg/kg)
Yangtze	0.048–0.063	890–954	7.1–10.2	83–110
Mingjiang	0.057	1010	35.7	76
Tuojiang	0.262	1130	82.8	85

One year after impoundment of the Three Gorges Dam, the area of the water reservoir expanded; the water level rose, the flow rate slowed down, the turbulent diffusion capacity of the water diminished. Lots of nutrients from flooded soil were dissolved in the water because of the long life of the pollutants. As result, the concentrations of nitrogen and phosphorus increased (Chai et al., 2009). As shown in Table 2.2, the mean concentrations of TN and TP before impoundment of the Three Gorges Dam were 1.461, 0.131 mg/L, respectively. One year after impoundment, the concentrations of TN and TP increased to 1.99 and 0.175 mg/L. This might be due to the fact that the water level rose, the flow rate slowed down, the turbulent diffusion capacity of the water diminished. The concentrations of TN and TP were 1.699 and 0.159 mg/L 1.5 years after impoundment, which were lower than those 1 year after impoundment (Chai et al., 2009).

The construction of the Three Gorges Reservoir will have a profound impact on the environment. The influence on the ecology and environment caused by the Three Gorges Project has been one of the hot issues of common concern at home and abroad. After the Three Gorges Reservoir was finished, the flow rate in typi-

Table 2.2 The concentrations of TN and TP in the main stream of the Yangtze River before and after impoundment of the Three Gorges Dam (Chai et al., 2009)

Sites	Before impoundment		1 year after impoundment		1.5 years after impoundment	
	TN	TP	TN	TP	TN	TP
The main stream	1.461	0.131	1.99	0.175	1.699	0.159

cal tributaries backwater area decreased, and the transparency increased. In the appropriate climatic conditions and nutrient conditions, there exists a potential danger of eutrophication. In this section, the distribution of nitrogen and phosphorus at the influx of the four main branches (Cuntan, Xiaojiang, Daning and Xiangxi Rivers) of the Three Gorges Reservoir on the Yangtze River were studied.

2.3 Distribution of Nitrogen

2.3.1 Water and Sediment Characteristics of Sample Sites

2.3.1.1 Sampling and Analyses

According to the pollution history survey and spot investigation, eutrophication would occur easily in the backwater zone of a reservoir. Therefore, the sample sites were selected in the main tributaries of the Three Gorges Reservoir— Cuntan, Xiaojiang, Daning and Xiangxi in 2006 and backwater zones of Xiaojiang, Daning, Xiangxi and the 5 sites along the Daning in 2007 using GPS (Fig. 2.1; Wang et al., 2009).

Cuntan, which is in Chongqing city, is located at the back of the Three Gorges Reservoir. Cuntan Port, the largest container terminal in the southwest, is here. A large amount of industrial wastewater and sewage was discharged here. Before the Three Gorges Reservoir was constructed, less sediment was deposited at Cuntan. However, after the dam was constructed, the flow velocity decreased and lots of sediments were deposited there. All of this would increase the risk of eutrophication.

Xiaojiang, which is the second largest tributary of the Yangtze River, has a length of 138.25 km and drainage area of 3043 km^2. It is an important protective area of the reservoir. Because of the construction of the Three Gorges Reservoir, the lands near Xiaojiang were flooded, and the flow rate in backwater areas was slow. There existed a higher risk of contamination.

The Daning River is a very important tributary of the Three Gorges Reservoir. It is also a famous scenic spot. Before the impoundment of the Three Gorges

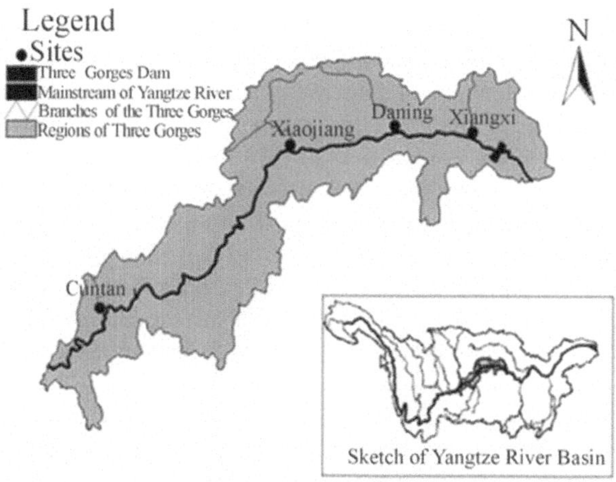

Fig. 2.1 The sampling map of backwater areas of main tributaries in the Three Gorges Reservoir

Reservoir, the Daning River was famous for its limpid water and precipitous gorge scenes. During water storage, the water quality of the 135-m backwater reach of the Daning River was monitored by the Chongqing Environmental Monitoring Center. It was found that the water color became green, dissolved oxygen (DO) in the upper water was oversaturated, and the nutritional state increased.

Xiangxi River is the largest tributary in Hubei of the Three Gorges Reservoir. The mainstream of the Xiangxi River is 94 km length, and has a total basin area of 3099 km^2 (Li et al., 2008). The Xiangxi River basin, which is rich in mineral resources, is one of China's three major phosphate-rich regions. Before the impoundment of the Three Gorges Reservoir, the main stream of the Xiangxi River with a mean annual flow of 65.5 m^3/s possesses the dilution capacity for the pollutants discharged into the river. After the impoundment of the Three Gorges Reservoir, the hydrological conditions and the water environment of the Xiangxi River downstream have been changed. The water level has risen and the flow rate slowed. This is not good for the spread of pollutants.

Surface water and bottom water (0–5 cm above the sampling sediments) were collected using sterile Niskin bottles. Sediments were collected with a Van Veen stainless steel grab sampler (Eijkelamp, Netherlands). Only the top 5 cm of sediments were used and mixed up in a polyethylene sealing bag, stored at 4 °C prior to conservation.

The pore water was obtained from sediments using a high speed centrifuge. In the laboratory, sediments were defrosted and air-dried at room temperature. The samples were then ground with agate mortar and pestal, sieved with a 60-mesh nylon sieve. At the same time, part of the surface water and bottom water samples were acidized, and part of the samples were filtrated by 0.45 μm cellulose acetate membrane. All of the sample collections were swing washed with surface water from sample sites three times to avoid pollution from the vessels. The temperature, DO, pH, conductivity and turbidity of the water were measured in situ.

Particle size analysis was performed by X-ray sedigraphy using a SALD-3001 Particle Analyzer (0.269–2000 m, R.S.D.<3%) (Shimadzu, Japan). Total organic carbon (TOC) content of sediments was determined using a Liqui TOC (Elementar, Germany) analyzer. Loss on ignition (LOI) measurement was based on weight losses after combustion at 550 °C. Total nitrogen (TN) and total phosphorus (TP) of sediments were analyzed by colorimetry after the digestion. Phosphate (PO_4^{3-}) was determined using the molybdenum-blue complex method (Kristina et al., 2005).

The water environment characteristics of sample sites are shown in Table 2.3. The range of dissolved oxygen content in the surface water of the tributaries from the Three Gorges Reservoir was 5.4–7.4 mg/L in autumn and 7.6–9.3 mg/L in spring. Therefore, an aerobic state dominated the surface water of the sample site in spring and autumn. The pH of the surface water ranged from 7.57 to 8.05 (with mean level of 7.85) in autumn and from 7.38 to 8.61 (with mean level of 8.10) in spring. The surface water from the Three Gorges Reservoir showed weak alkaline generally. In 2007, the surface water and the overlying water were both sampled. Compared with the surface water, the levels of pH (7.02–8.36) and dissolved oxygen content (5.15–8.67) were decreased in the overlying water.

2.3.1.2 Sediment Properties

As reported in the literature, the sediment properties and composition (such as organic matter and metal oxide) would influence the sediment adsorption capacity of phosphorus. The main properties of the sediment samples were shown in Table 2.4.

In sediment sampling in 2006, organic matter, expressed as TOC and LOI, showed the lowest value of 0.39% and 2.92%, respectively, in Cuntan sediment, the highest of 1.17% and 6.82% in Xiaojiang sediment. TP contents in different sediments were in decreasing order: Xiaojiang>Cuntan>Xiangxi>Daning. That is, the Xiaojiang sediment with the highest content of organic matter exhibited a maximum capacity for phosphorus retention, while the sediment possessing the lowest content of organic matter (Cuntan) did not show the minimum capacity for phosphorus retention, but was higher than Xiangxi and Daning sediments. This result indicated that organic matter increased phosphorus retention in sediments; however, there must be other properties influencing the adsorption of phosphorus. The content of metal oxides (Fe and Al) has been considered to be a main factor that determines phosphorus retention because of the high specific surface of the hydroxides. In our study, the sum of Fe and Al followed the order of Xiaojiang> Cuntan>Daning>Xiangxi, which may be the reason that Cuntan showed a higher retention capacity for phosphorus than Daning and Xiangxi. Another factor affecting phosphorus retention is the grain size. It is noticeable that the sediments used in this study were primarily composed of clay and silt-sized particles with few sand-sized particles. The Xiaojiang sediment showed the highest clay content and at the same time possessed maximum values of contents of organic matter, TN and TP, indicating that the increase in the clay fraction with the larger specific area accelerated the adsorption of the pollutant.

The organic matter of sediment plays an important role in migration and release of pollutants. The mineralization process of organic matter, which consumes oxygen and releases nutrients such as C, N, P, S and so on, will cause serious deterioration of the water and eutrophication. The organic matter plays an important role in the migration behavior of the nutrient substances, heavy metals and organic compounds in sediment by adsorption and complexation. Therefore, organic matter plays an important role in the environmental chemistry and pollution chemistry of sediments. TOC and LOI, which are shown in Fig. 2.2, were used to characterize the organic matter content in sediment. In this study, the range of TOC was 0.39%–1.17% in sediments sampled in autumn, and0.58%–2.19% in spring. And the ranges of LOI were 2.92%–6.82% and 2.63%–7.23%, respectively. The values of LOI in all samples were higher than TOC. The result indicated that heating oxidation was stronger than chemical oxidation. A higher level of LOI than TOC might be due to the fact that the LOI

Table 2.3 The sample sites and the environmental characteristics at main tributaries of the Three Gorges Reservoir

Sample time		Position	E	N	pH	Conductivity (μm/cm)	DO (mg/L)	Eh (mV)	Temp. (℃)	Turbid (NTU)	Chla (mg/L)
2006.10	The surface water	Cuntan	106°37’	29°37’	8.05	353	7.46	55.8	22.6	125.0	2.26
		Xiaojiang	108°39’	30°57’	7.57	398	5.40	78.2	21.7	7.0	8.75
		Daning	109°54’	31°08’	7.98	457	6.80	88.1	22.1	7.0	2.40
		Xiangxi	110°46’	30°58’	7.81	401	7.02	83.4	22.1	3.0	3.01
2007.4	The surface water	Xiaojiang	108°40’	30°57’	7.38	471	7.75	130.2	19.5	1.2	6.66
		Daning A	109°52’	31°11’	7.6	457	7.55	120.7	20.5	2.0	1.03
		Daning B	109°54’	31°09’	8.17	497	8.29	120.4	21.0	2.5	1.51
		Daning C	109°54’	31°08’	8.37	491	9.25	70.8	21.5	0.8	0.80
		Daning D	109°53’	31°07’	8.18	485	8.05	128.4	20.0	1.0	1.23
		Daning E	109°54’	31°05’	8.4	409	9.52	80.2	19.8	2.1	1.09
		Xiangxi	110°46’	30°58’	8.61	433	7.81	44.1	19.5	2.6	12.14
	The overlying water	Xiaojiang	108°40’	30°57’	7.02	468	5.15	119.4	17.5	1.3	
		Daning A	109°52’	31°11’	7.41	455	5.20	121.1	16.5	0.8	
		Daning B	109°54’	31°09’	7.95	491	6.96	105.8	16.5	1.0	
		Daning C	109°54’	31°08’	8.18	504	6.90	61.3	17.0	1.3	
		Daning D	109°53’	31°07’	8.08	487	7.60	135.8	17.5	1.5	
		Daning E	109°54’	31°05’	8.32	463	8.67	72.3	16.0	3.2	
		Xiangxi	110°46’	30°58’	8.36	457	7.63	24.8	15.5	2.3	

Table 2.4 The sediment properties of the main tributaries of the Three Gorges Reservoir

Sample time	Sites	TOC (%)	LOI (%)	TN (mg/g)	TP (mg/g)	Fe (mg/g)	Al (mg/g)	Ca (mg/g)	Clay (%)	Silt (%)	Sand (%)	P/(Fe+Al) (μmol/mmol)
2006.10	Cuntan	0.39	2.92	0.14	0.83	31.44	47.29	47.60	23.25	43.76	32.99	11.58
	Xiaojiang	1.17	6.82	0.56	0.95	43.32	54.78	38.65	53.82	46.17	0.01	10.94
	Daning	0.51	5.75	0.27	0.75	23.63	7.75	101.32	45.89	53.02	1.19	34.12
	Xiangxi	0.96	4.98	0.11	0.76	25.31	4.75	1.66	36.29	53.25	10.46	39.05
2007.4	Xiaojiang	0.76	2.63		1.00	27.61	49.60	8.27	29.48	50.74	19.79	13.87
	Daning A	0.96	4.31		0.83	40.50	69.09	17.64	50.07	48.03	1.90	8.13
	Daning B	2.19	7.23		1.18	38.64	50.66	101.92	34.70	57.79	7.52	14.88
	Daning C	0.58	3.85		0.90	24.89	47.78	137.62	37.47	59.50	3.03	13.16
	Daning D	1.54	6.19		0.42	37.31	66.40	16.04	44.51	52.66	2.83	4.29
	Daning E	0.72	4.07		0.44	37.31	65.57	69.81	44.27	52.21	3.52	4.61
	Xiangxi	0.94	4.25		1.17	29.99	43.28	2.56	37.78	52.94	9.28	17.63

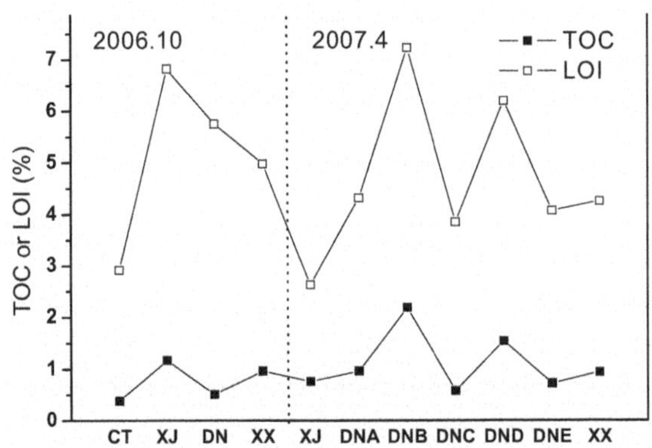

Fig. 2.2 TOC and LOI in sediments of main tributaries of the Three Gorges Reservoir in 2006 and 2007 (CT: Cuntan; XJ: Xiaojiang; DN: Daning; XX: Xiangxi)

value included water from sediment and crystal water.

As known, the sediment comprises different sizes of particles. Because of different specific surface areas and different weights, different size particles have different characteristics and different effects in the exchange of nutrients (nitrogen and phosphorus) and other pollutants in the solid-liquid interface. In Table 2.4, the amount of clay and silt is more than sand in the surface sediments. The range of clay levels in the sediments was 23.25%–53.82% in autumn and 29.48%–50.07% in spring, respectively.

2.3.2 Distribution of Nitrogen

Water samples were filtered through a polycarbonate filter (0.45 mm pore-size) for chemical analysis. Total nitrogen (TN) and total dissolved nitrogen (TDN) in water sediments were analyzed by a potassium sulfate oxidation ultraviolet spectrophotometry method. Nitrate nitrogen (NO_3^--N) was determined by a hydrochloric acid ultraviolet spectrophotometry method. Nitrite nitrogen (NO_2^--N) was determined by an N-(1-naphthyl) ethylenediamine spectrophotometry method. Ammonium nitrogen (NH_4^+-N) was determined by a salicylic acid method.

In Table 2.5, the concentration of total nitrogen (TN), total dissolved nitrogen (TDN), nitrate nitrogen (NO_3^--N), nitrite nitrogen (NO_2^--N) and ammonia nitrogen (NH_4^+-N) in the surface water, overlying water and pore water from the Three Gorges Reservoir tributaries are listed. As shown, the ranges of TN, TDN, NO_3^--N, NO_2^--N and NH_4^+-N concentration were 3.615–7.265, 1.487–6.844, 0–0.34, 0–0.007 and 0–0.011 mg/L in autumn, and were 1.048–3.807, 0.744–1.510, 0.1–0.7, 0–0.007 and 0–0.16 mg/L in spring, respectively. The concentrations of different speciation of nitrogen in the pore water from most sample sites were higher than in the surface water and overlying water. In addition, at most sample sites, the soluble inorganic nitrogen (NO_3^--N, NO_2^--N and NH_4^+-N) levels were far lower than that the TDN levels. This result indicated that soluble organic nitrogen is the dominating form of nitrogen in the tributaries of the Three Gorges Reservoir.

As known, the Xiangxi and Daning Rivers are the main tributaries of the Three Gorges Reservoir. The Xiangxi is the largest tributary in Hubei Reservoir of the Three Gorges Reservoir (Li et al., 2008). And the Daning River is a very important tributary in the Three Gorges Reservoir. It was famous for its limpid water and precipitous gorge scenes before the impoundment of the Three Gorges Reservoir (Zhong et al., 2005). The concentrations of TN and the flux dynamics of nitrogen in the water of the Xiangxi and Daning Rivers were reported by other researchers (Li et al., 2008; Zhong et al., 2005). Based on monitoring data for the Xiangxi River from September 2000 to June 2005 (Li et al., 2008), Xiangxi Bay receives 1623.49 t of total nitrogen (TN) annually, and Xiangxi River accounts for 68.50% of the total nitrogen fluxes of Xiangxi Bay. In the Xiangxi River, dissolved inorganic nitrogen (DIN) is the dominating form of nitrogen. The fluxes of DIN and TN are high during summer, mid-spring and autumn, and relatively low in winter. Based on monitoring data in the 135-m backwater reach of the Daning River in 2003, Zhong et al. (2005) reported that about 711.8 t TN were input. About 339.4 t TN were input from water flows and 372.2 t TN were retained in the backwater reach. About 244.8 t TN was deposited in the backwater and about 90.2 t TN was absorbed by algae.

Table 2.5 The concentrations of nitrogen in water in the main tributaries of the Three Gorges Reservoir

Sample time		Position	TN (mg/L)	TDN (mg/L)	NO_3^--N (mg/L)	NO_2^--N (mg/L)	NH_4^+-N (mg/L)
2006.10	Surface water	Cuntan	4.430	2.365	0.000	0.000	0.011
		Xiaojiang	3.615	2.400	0.000	0.007	0.000
		Daning	6.790	1.487	0.340	0.000	0.011
		Xiangxi	7.265	6.844	0.280	0.000	0.000
	Pore water	Cuntan		8.213	0.241	0.009	0.039
		Xiaojiang		8.440	4.300	0.018	0.360
		Daning		7.719	0.100	0.024	0.020
		Xiangxi		9.790	4.600	0.050	0.100
2007.4	Surface water	Xiaojiang	2.348	0.764	0.550	0.040	0.160
		Daning A	1.048	0.757	0.300	0.000	0.010
		Daning B	1.772	1.068	0.400	0.000	0.000
		Daning C	2.025	0.744	0.100	0.000	0.000
		Daning D	3.807	0.974	0.200	0.000	0.000
		Daning E	2.294	1.510	0.100	0.010	0.030
		Xiangxi	3.432	1.218	0.700	0.000	0.000
	Overlying water	Xiaojiang	0.724	0.723	0.600	0.010	0.050
		Daning A	0.734	0.734	0.600	0.000	0.000
		Daning B	1.393	1.341	0.200	0.000	0.000
		Daning C	0.920	0.902	0.600	0.000	0.060
		Daning D	1.628	1.243	0.800	0.000	0.070
		Daning E	1.763	1.596	0.700	0.020	0.050
		Xiangxi	2.071	1.462	0.800	0.000	0.000
	Pore water	Xiaojiang		8.775	3.600	0.010	0.190
		Daning A		12.405	4.700	0.000	0.220
		Daning B		4.613	2.100	0.000	0.113
		Daning C		3.552	2.400	0.010	0.120
		Daning D		1.264	2.500	0.000	0.145
		Daning E		1.412	1.900	0.000	0.155
		Xiangxi		1.808	0.900	0.000	0.310

There are more than 30 lakes in southwestern China. Most of these lakes have been contaminated by human activities and they are experiencing a serious deterioration in the water quality. The increase of nutrient concentration is an important factor in these problems. Research was conducted on the natural wetland on Bogong Island in West Wulihu Lake Bay, located in the north-west of

Taihu Lake (Yang et al., 2007). In Table 2.6, the mean concentrations of TN, NO_3^--N, NH_4^+-N and DON in a solution of soil from the natural wetland of Taihu Lake were 2.46, 0.98, 0.08 and 1.35 mg/L, respectively (Yang et al., 2007). The mean concentration of TN was higher than the eutrophication level. Among TN, the concentration of DON dominated the distribution of N. And as Table 2.6 shows, the levels of TN and DON in Taihu were highest in winter. Lake Lugu, Lake Erhai and Lake Baihua are in southwestern China. Lake Lugu, which is regarded as one of the best naturally preserved lakes in China, is isolated from industrial areas due to its high altitude. However, Lake Erhai and Lake Baihua are both polluted lakes. The concentrations of N here were the highest in this region (Wu et al., 2001). In Table 2.6, the mean concentrations of TN in Lake Lugu, Lake Erhai and , Lake Baihua were 0.26, 0.07 and 1.26 mg/L in overlying water and 1.27, 1.11 and 1.22 mg/L in pore water, respectively (Wu et al., 2001). And as reported, among the three dissolved N forms (NH_4^+, NO_3^- and NO_2^-), NO_3^- was dominant in the overlying water, while NH_4^+ was dominant in the pore water, suggesting that strong regeneration mainly occurred near the sediment/water interface (Wu et al., 2001).

Table 2.6 The concentrations of TN, NO_3^-, NH_4^+ and DON in soil solution in aquatic-terrestrial ecotone of Taihu Lake (Yang et al., 2007), and the concentration of TN in the overlying water and pore water of lakes Lugu, Erhai and Baihua (Wu et al., 2001)

		TN (mg/L)	NO_3^- (mg/L)	NH_4^+ (mg/L)	DON (mg/L)
Taihu	Spring	2.79	0.71	0.11	1.76
	Summer	2.14	0.95	0.02	1.21
	Autumn	2.01	1.57	0.05	0.45
	Winter	2.90	0.70	0.13	1.97
	Mean	2.46	0.98	0.08	1.35
Overlying water	Lake Lugu	0.26			
	Lake Erhai	0.07			
	Lake Baihua	1.26			
Pore water	Lake Lugu	1.27			
	Lake Erhai	1.11			
	Lake Baihua	1.22			

For two anthropogenically influenced estuarine systems—the Yealm and Plym estuaries in southwest England (El-Sayed et al., 2008), the concentrations of DON in both estuaries were generally$<$80 μmol/L. DON showed non-conservative distributions, resulting from external and internal inputs and in situ reactivity. DON contributed 38%±22% and 36%±17% to the TDN (El-Sayed et al., 2008). DON was a larger fraction of the TDN during the summer and autumn relative to winter and spring, indicating the influence of bacterioplankton release on nitrogen cycling in the estuaries. As reported, the relationship of the spatial distribution of C and N is important for understanding the biogeochemical processing. As known, the Bengal basin is one of the geologically youngest and tectonically most active drainage regimes in the world and includes the total Lower Ganges-Brahmaputra-Meghna (GBM) river system, one of the highest sediment dispersal systems in the world. In the GBM river system (Datta et al., 1999), variations in total nitrogen (TN) contents in the bulk and <63 mm fraction of the bed sediments are statistically insignificant, but total carbon (TC) content varies significantly among the three sub-basins. Although excellent correlation between C_{org} and TN suggests their co-origin, the influence of inorganic nitrogen is evident in very low C/N ratios. The C/N ratio varies from 2 to 11, suggesting complete degradation of the organic matter and/or enrichment of inorganic nitrogen in the sediments (Datta et al., 1999).

2.4 Distribution of Phosphorus

2.4.1 Distribution of Phosphorus

Water samples were filtered through a polycarbonate filter (0.45 mm pore-size) for chemical analysis. The sediment samples were digested for analysis. Total phosphorus (TP) and total dissolved phosphorus (TDP) were analyzed by colorimetry. Phosphate (PO_4^{3-}-P) was determined using the molybdenum-blue complex method.

In Table 2.7, the concentrations of total phosphorus (TP), total dissolved phosphorus (TDP) and soluble phosphate (PO_4^{3-}-P) in the surface water, overlying water and pore water of all sampling sites from the Three Gorges Reservoir were shown. Among them, TP included all forms of phosphorus (inorganic phosphorus, particulate phosphorus, organic phosphorus and so on) in water. TDP included dissoluble organic phosphorus and inorganic phosphorus. PO_4^{3-}-P included orthophosphate and a small amount of hydrolysis condensed phosphate in the water samples with filtering. As shown, the concentrations of TP in the surface water ranged from 0.077 to 0.139 mg/L in autumn and from 0.057 to 0.136 mg/L in spring. And the range of TDP and PO_4^{3-}-P concentrations was 0.052–0.106 and 0.033–0.082 mg/L in autumn and 0.032–0.125 and 0.005–0.063 mg/L in spring. Therefore, the concentrations of phosphorus in most samples in autumn were similar to those in spring. In Figs. 2.3 and 2.4, the concentrations of TDP and PO_4^{3-}-P in the pore water were higher than that in the surface water and overlying water (Wang Y et al., 1999).

As known, the Xiangxi River is the largest tributary of Hubei Reservoir within the Three Gorges Reservoir. The concentrations of TP and the flux dynamics of phosphorus in the water of the Xiangxi and its main tributary the Gaolan were also reported by other researchers (Li et al., 2008). Based on the monitoring data for the Xiangxi from September 2000 to June 2005 (Li et al., 2008), Xiangxi Bay receives 331.85 tons of total phosphorus annually, and Xiangxi River accounts for 91.74% of the total phosphorus fluxes of Xiangxi Bay. Orthophosphate is the

Table 2.7 The concentrations of phosphorus in water from the main tributaries of the Three Gorges Reservoir

Sample time		Position	TP (mg/L)	TDP (mg/L)	PO_4^{3-}-P (mg/L)
2006.10	Surface water	Cuntan	0.107	0.106	0.036
		Xiaojiang	0.083	0.052	0.050
		Daning	0.077	0.060	0.033
		Xiangxi	0.139	0.083	0.082
	Pore water	Cuntan	/	0.846	0.691
		Xiaojiang	/	0.327	0.281
		Daning	/	0.218	0.130
		Xiangxi	/	1.070	0.938
2007.4	Surface water	Xiaojiang	0.136	0.125	0.063
		Daning A	0.057	0.037	0.014
		Daning B	0.078	0.071	0.005
		Daning C	0.077	0.073	0.007
		Daning D	0.089	0.032	0.006
		Daning E	0.062	0.045	0.020
		Xiangxi	0.103	0.056	0.015
	Overlying water	Xiaojiang	0.057	0.056	0.043
		Daning A	0.045	0.032	0.029
		Daning B	0.050	0.025	0.021
		Daning C	0.053	0.051	0.028
		Daning D	0.073	0.056	0.040
		Daning E	0.084	0.048	0.024
		Xiangxi	0.110	0.104	0.065
	Pore water	Xiaojiang	/	0.439	0.328
		Daning A	/	0.450	0.150
		Daning B	/	0.391	0.249
		Daning C	/	0.221	0.159
		Daning D	/	0.284	0.149
		Daning E	/	0.236	0.137
		Xiangxi	/	0.620	0.448

dominant form of phosphorus in the Xiangxi, and is relatively low in its main tributary, the Gaolan. The fluxes of phosphorus are high during summer and late spring, relatively low during winter and late autumn in the Gaolan, but fluctuate irregularly in the Xiangxi. The phosphorus in the Gaolan is mainly caused by non-point source pollutants, while point source pollutants of phosphorus play an important role there. The Daning River is a very important tributary in the Three Gorges Reservoir. It was famous for its limpid water and precipitous gorge scenes before the impoundment of the Three Gorges Reservoir (Zhong et al., 2005). Eutrophication monitoring in the 135-m backwater reach of the Daning River in

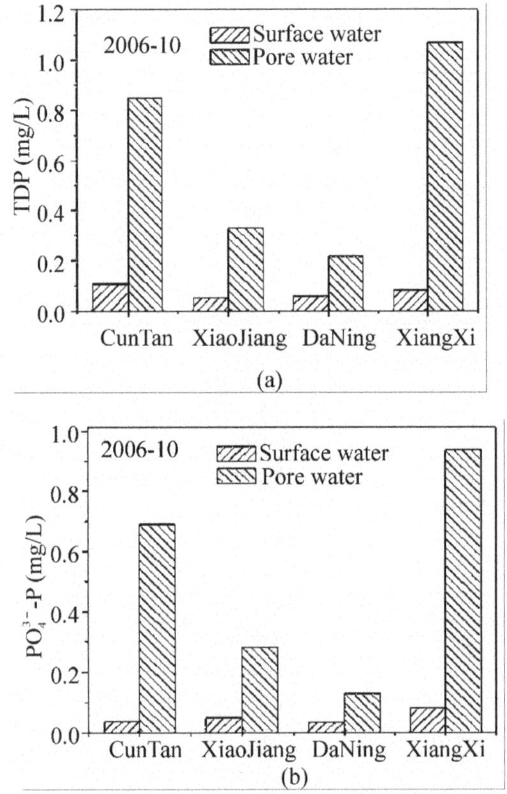

Fig. 2.3 The concentrations of TDP (a) and PO_4^{3-}-P (b) in the surface water and pore water from the main tributaries of the Three Gorges Reservoir in 2006

2003 was conducted (Zhong et al., 2005). During water storage, about 27.6 t TP were input. About 11.3 t TP were input from water flows and 16.3 t TP were retained in the backwater reach. The main sources of nutrients are runoff from farmland during storage, background input from the upper reaches and saturated soil in flooded fields. The concentration of TP in Daning River was 0.03 mg/L. After impoundment of the Three Gorges Dam (2003), the area of the reservoir expanded; the water level rose; the flow rate slowed down; the turbulent diffusion capacity of water diminished. Therefore, lots of phosphorus from flooded soil is dissolved in water because of the long life of the pollutants.

As known, most of the lakes in southwestern China have been contaminated by human activities and they are experiencing a serious deterioration in the water quality. In the last two decades, approximately one billion dollars were spent in the restoration of the lake environment in China. Investigation of the distribution of the nutritive element in lakes was needed. Research was conducted on the natural wetlands on Bogong Island in West Wulihu Lake Bay, located in the north-west of Taihu Lake (Yang et al., 2007). In Table 2.8, the mean concentration of TP was 0.057 mg/L in solution of soil from the natural wetlands of Taihu Lake

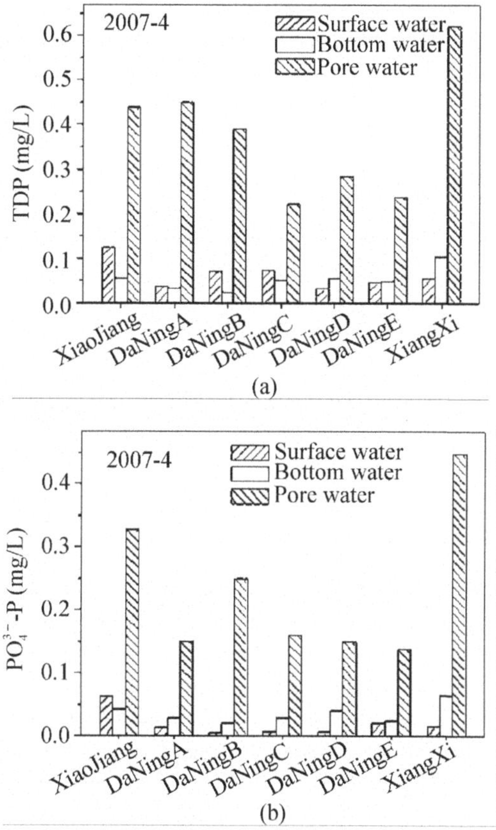

Fig. 2.4 The concentrations of TDP (a) and PO_4^{3-}-P (b) in the surface water and pore water from the main tributaries of the Three Gorges Reservoir in 2007

(Yang et al., 2007), which was higher than the eutrophication level for P (0.02 mg/L). That indicated that there was the potential risk of eutrophication. And as Table 2.8 shows, the levels of TP were highest in winter and lowest in spring. Lake Lugu, Lake Erhai and Lake Baihua are in southwestern China. Lake Erhai and Lake Baihua are two polluted lakes. However, Lake Lugu, which is regarded as one of the best naturally preserved lakes of China, is isolated from industrial areas due to its high altitude. The concentrations of TP in the overlying water were below 0.01 mg/L in these three lakes, but they rapidly increased near the interface, and were much higher in the pore water. The increase of TP in the pore water could be attributed to the release of TP due to the dissolution of iron oxides (Sundby et al., 1992), or attributed to the decomposition of organic matter in the sediments.

Table 2.8 The concentrations of TP in soil solution in aquatic-terrestrial ecotone of Taihu Lake (Yang et al., 2007), and in the overlying water and pore water of lakes Lugu, Erhai and Baihua (Wu et al., 2001)

Site		TP (mg/L)
Taihu	Spring	0.025
	Summer	0.057
	Autumn	0.063
	Winter	0.082
	Mean	0.057
Lake Lugu	Overlying water	<0.0017
Lake Erhai		<0.0017
Lake Baihua		0.0084
Lake Lugu	Pore water	<0.01
Lake Erhai		0.07
Lake Baihua		0.15

In the sediments from Tomales Bay (Vink et al., 1997), most of the phosphorus (70%–80%) contained in suspended and deposited sediment samples was found in the organic and residual phases. Pore water from sediment cores was analyzed for phosphate. Within the Tomales Bay sediments, pore water was depleted in phosphate relative to dissolved inorganic carbon and ammonium, suggesting that phosphate released from organic matter decomposition is being removed from the pore water (Vink et al., 1997).

2.4.2 Speciation Analysis of Phosphorus

2.4.3.1 Method

An analytical protocol—the SMT protocol for the determination of the extractable phosphorus content in the sediments has been harmonized through laboratory studies in the frame of the Standards Measurements and Testing Program of the European Commission. The SMT protocol provides useful tools in the field of water environment management, especially at a time when quality assurance and data comparability are of paramount importance in laboratory analysis (Ruban et al., 2001). Moreover, the EDTA method of the extractable phosphorus content in the sediments was used tor more effect (Golterman, 1996). These two protocols were listed.

● *SMT protocol*

The protocol consisted of the three extraction procedures that were applied to 0.2 g aliquots of sediment samples:

1) An extraction (16 h) using 20 mL of 1 mol/L NaOH was performed and after centrifugation and separation of the supernatant liquid, the residue was extracted again with 1 mol/L HCl (16 h). Fe/Al-P was determined in the extract. 1 mol/L HCl was added to one aliquot and left to stand for 16 h to precipitate organic matter. Ca-P was determined in the supernatant liquid.

2) An extraction (16 h) with 20 mL of 1 mol/L HCl was performed to determine IP. The residue of this extraction was placed in a porcelain crucible and calcined in a furnace for 3 h at 450 °C. Then, the residue was extracted again (16 h) with 20 mL of 1 mol/L HCl. After centrifugation, OP was determined in the extract.

3) After sample calcination during 3 h at 450 °C, a single extraction (16 h) with 20 mL of 3.5 mol/L HCl was carried out, and TP determined in the extract.

Detailed experimental conditions are described in Schedule 2.1.

● *EDTA Protocol*

The following extractants (Hieltjes and Lijklema, 1980; Schedule 2.2) are used:

1) Ca-EDTA, 0.05 mol/L dissolve 18.6 g of $Na_2EDTA{\cdot}2H_2O$ (Titriplex III) together with 7.35 g of $CaCl_2{\cdot}2H_2O$ in 1l H_2O '$Ca(Cl_2)$-EDTA'. Add Trisbuffer till pH=9 (about 13 g). After the addition of 1% Na-dithionite, just before extraction, the pH value must be 7–8. In one experiment $CaCO_3$ was used: '$Ca(CO_3)$-EDTA' .

2) Na_2-EDTA, 0.1 mol/L dissolve 37.2 g of $Na_2H_2EDTA{\cdot}2H_2O$ in 1 L of H_2O. The pH value is about 4.5 (Golterman, 1996).

2.4.3.2 Distribution

The contents and speciation of phosphorus are different in the different sediments. The study of the speciation of phosphorus would be the foundation for discussing the transfer and transformation of phosphorus in the sediments-water interface.

In Table 2.9, the concentration of TP in the sediments ranged from 0.746 to 0.946 mg/g in autumn and from 0.420 to 1.180 mg/g in spring. The range of Inorg-P concentration was 0.453–0.663 mg/g in autumn and 0.184–0.699 mg/g in spring, which accounts for 59.29%–78.82% and 42.05%–62.30% of TP, respectively. The range of Org-P concentration was 0.168–0.283 mg/g in autumn and 0.170–0.478 mg/g in spring, which accounts for 20.22%–36.39% and 29.90%–45.48% of TP, respectively. The concentration of Inorg-P was higher than Org-P in all sediments except Daning D sediment. As known, the Inorg-P includes Fe/Al-P and Ca-P. The ratio values of Fe/Al-P and Ca-P (0.033–0.788 mg/g) were less than 1 in all samples except Xiangxi sediment sampled in 2007, which indicated that Inorg-P was dominated by Ca-P. As reported, the high volume accumulation of Ca-P would be related to the increase in the eutrophication level (Bostrom et al., 1982). Org-P could be divided into acid extractable organic phosphate and alkaline extractable organic phosphate. Mineralization or degradation of Org-P compounds could play an important role in the process of diffusion of SRP from sediments to pore water. In some certain conditions, the Org-P_{ac} could hydrolyze or mineralize into SRP which has potential bioavailability and is more related to eutrophication. And Org-$P_{alk,}$ which has soluble fulvic acid components, has comparatively weak bioavailability. In the most samples, the concentration of Org-P_{ac} was higher than Org-P_{alk}, which indicated that the potential bioavailability of phosphorus in the sediments was high.

Downstream on the Yangtze River, before the construction of the Three Gorges Dam, the phosphorus fractionation in the total suspended sediment of the Yangtze at Datong (Duan et al., 2008) was dominated by Ca-P (66%), followed by occluded ferrous-P (22%), Al-P (6%) and Fe-P (4%), occluded Al-P and adsorbed P (<1%). The contents of adsorbed P, occluded Fe-P, Al-P, Fe-P and PBAP (total phosphorus and particulate bioavailable-P), together with total N and OC, were all

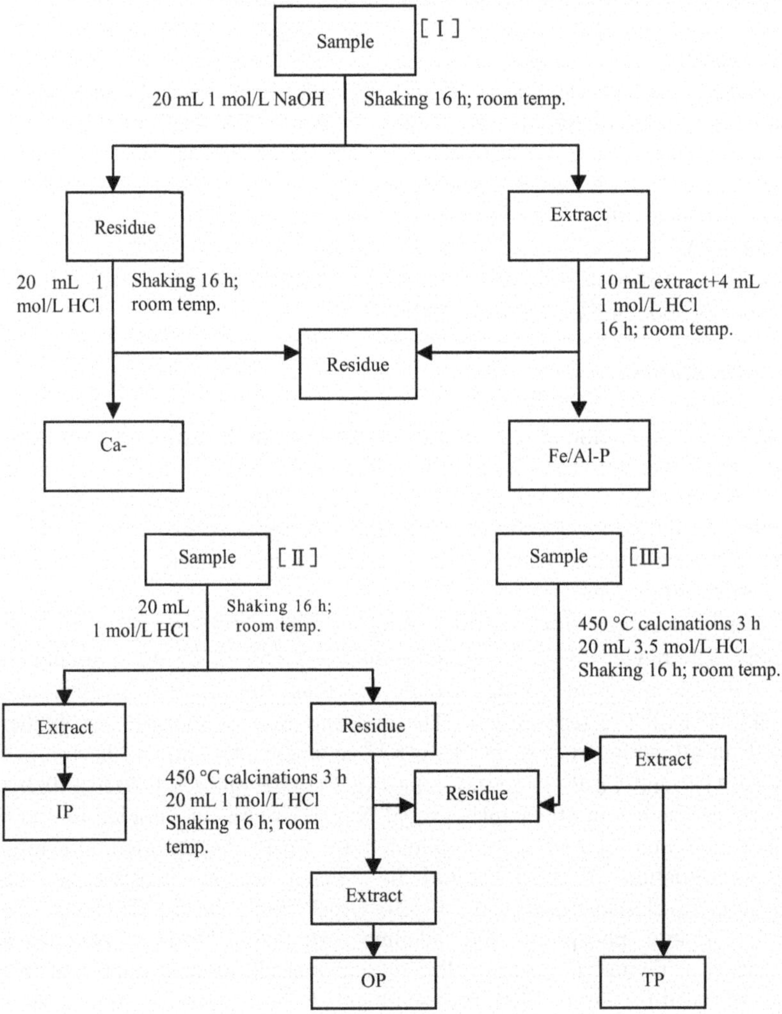

Schedule 2.1 Detailed experimental conditions of SMT protocol

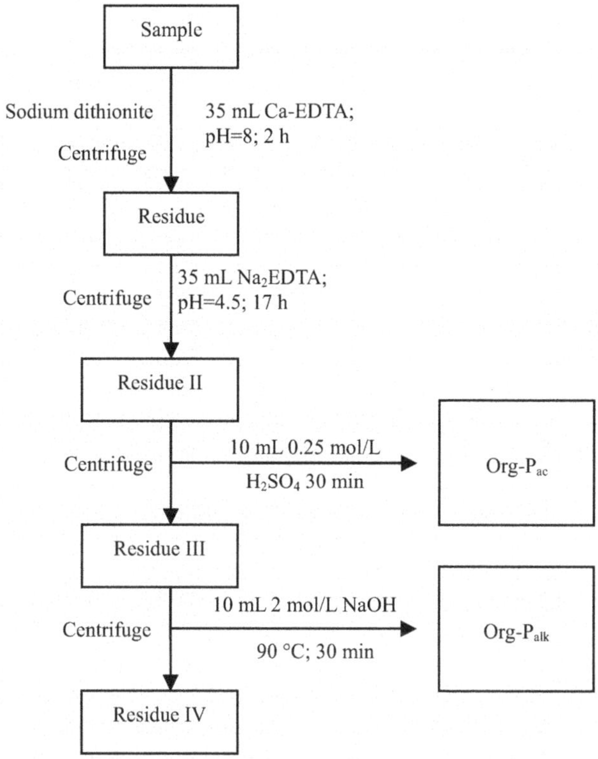

Schedule 2.2 Detailed experimental conditions of EDTA protocol

Table 2.9 The concentrations of different speciations of phosphorus in sediments samples from the main tributaries of the Three Gorges Reservoir

Sample time	Sample sites	TP (mg/g)	Org-P (mg/g)	Inorg-P (mg/g)	Fe/Al-P (mg/g)	Ca-P (mg/g)	Org-P_{ac} (mg/g)	Org-P_{alk} (mg/g)
2006.10	Cuntan	0.831	0.168	0.655	0.018	0.543	0.143	0.028
	Xiaojiang	0.946	0.283	0.663	0.032	0.492	0.173	0.115
	Daning	0.746	0.252	0.532	0.190	0.247	0.194	0.053
	Xiangxi	0.764	0.278	0.453	0.116	0.224	0.075	0.168
2007.4	Xiaojiang	1.000	0.299	0.623	0.152	0.402	0.177	0.106
	Daning A	0.830	0.320	0.349	0.074	0.171	0.102	0.189
	Daning B	1.180	0.478	0.691	0.121	0.481	0.283	0.174
	Daning C	0.900	0.386	0.513	0.249	0.316	0.231	0.098
	Daning D	0.420	0.191	0.184	0.031	0.070	0.040	0.140
	Daning E	0.440	0.170	0.211	0.031	0.098	0.063	0.105
	Xiangxi	1.170	0.441	0.699	0.465	0.189	0.132	0.282

lowest during the summer flood and the highest during the base-flow period. The seasonal changes in the composition of P were likely a result of hydrodynamic sorting, whereby finer particles, which were enriched in organic matter and Al/Fe oxides generally (Spitzy and Ittekkot, 1991), were transported during the low-flow period, while particles were delivered during the summer flood due to the carrying capacity and large erosion. The minimal seasonality of Ca-P suggests that Ca-P was not affected by hydrodynamic sorting. Actually, Ca-P was positively correlated with $CaCO_3$, indicating that the majority of Ca-P consisted of phosphorus minerals that were in coexistence with calcite or aragonite. The uncommonly high content of Ca-P in the Yangtze TSS, compared to the Zhujiang and other rivers (Sutula et al., 2004), might be related to the shale, carbonate rocks and purple calcite sandstone in the upper drainage basin, which were the major source of total suspended sediment in the Yangtze (Xu et al., 2006). This conclusion was supported by the characteristically high Ca-P contents in the shale, the fluvial and soil sediments (Xiangxi River) of the upper stream (Fu et al., 2006).

The P fractionation in the estuarine sediments of the Daliao River showed that Ca-P, residual P, Al-P, reductant-soluble P, Fe-P, and soluble and loosely bound P were on average 44.5, 21.6, 13.6, 11.7, 8.9 and 0.2% of TP, respectively (Wang P et al., 2009). With the gradual increase in TP content, Al-P, reductant-soluble P, and Fe-P generally increased, while the rest of P did not. This might indicate that anthropogenic P is bound to Fe and Al oxides. Regression analysis showed that Al-P, the sum of Fe-P and reductant-soluble P, was correlated to the contents of total Al and Fe. Additionally, Ca-P was not correlated to the content of total Ca in the sediment, suggesting that Ca-P was mainly from antigenic marine origins.

The distribution and forms of phosphorus in Poyang Lake sediments were obtained by chemical extraction methods to understand the potential adsorption and release of phosphorus (Xiang and Zhou, 2011). The results showed that phosphorus in sediments consisted mainly of inorganic phosphorous species, including Fe-P, Al-P, Ca-P and dissolved phosphorus, as well as Org-P species. The distribution of P species in the surface sediments showed a decreasing trend from the outlet to entrance of Poyang Lake. The content of TP in the surface sediments ranged from 688.29 to 825.36 mg/kg and inorganic phosphorous was the major species. Fe-P, accounting for greater than 40% of the TP, was the dominant form of inorganic phosphorous, while Al-P and Ca-P accounted for more than 20%, and dissolved phosphorus was less than 5%. The content of Org-P in the surface sediments was lower, being less than 15%. TP was correlated with Fe-P, Ca-P, Ca-P and Org-P positively, while Org-P was correlated with Fe-P and Al-P positively, but only related to Ca-P weakly. In the core sediments, the contents of TP and forms of phosphorus obviously decreased with depth, while they were enriched in the surface sediments. The TP and forms of phosphorus contents were significantly higher in the upper layers and had a tendency to decrease toward the bottom of the core.

References

Beat M, Michael B, Yao ZP, Zhang XF, Wang D, August P (2008) How polluted is the Yangtze River? 0Water quality downstream from the Three Gorges Dam. Sci Total Environ 402:232-247.

Bostrom B, Janssco M, Forsberg C (1982) Phosphorus release from lake sediment. Hydrobiol 18:5-59.

Carpenter SR, Caraco NF, Correll DL, Howarth RW, Sharpley AN, Smith VH (1998) Nonpoint pollution of surface waters with phosphorus and nitrogen. Ecological Applications 8:559-568.

Cao ZJ, Zhang XB, Ai NS (2011) Effect of sediment on concentration of dissolved phosphorus in the Three Gorges Reservoir. Int J Sediment Res 26:87-95.

Cecilia L, Malin L, Mikaelvon N, Erik B (2005) A multivariate assessment of coastal eutrophication. Examples from the Gulf of Finland, northern Baltic Sea. Mar Pollut Bull 50:1185-1196.

Cecilia L, Britt-Marie J, Erik B (2009) The spreading of eutrophication in the eastern coast of the Gulf of Bothnia, northern Baltic Sea—An analysis in time and space. Estuar Coast Shelf Sci 82:152-160.

Chai C, Yu ZM, Shen ZL, Song XX, Cao XH, Yao Y (2009) Nutrient characteristics in the Yangtze River Estuary and the adjacent East China Sea before and after impoundment of the Three Gorges Dam. Sci Total Environ 407:4687-4695.

Datta DK, Guptab LP, Subramanian V (1999) Distribution of C, N and P in the sediments of the Ganges-Brahmaputra-Meghna river system in the Bengal basin. Org Geochem 30:5-82.

Duan SW, Liang T, Zhang S, Wang LJ, Zhang XM, Chen XB (2008) Seasonal changes in nitrogen and phosphorus transport in the lower Yangtze river before the construction of the Three Gorges Dam. Estuar Coast Shelf Sci 79:239-250.

El-Sayed AB, Alan DT, Eric PA (2008) Distributions and seasonal variability of dissolved organic nitrogen in two estuaries in SW England. Mar Chem 110:153-164.

Fu CY, Fang T, Deng N (2006) The research of phosphorus in Xiangxi river near the Three Gorges, China. Environ Geol 49:923-928.

Gao L, Li DJ, Ding PX (2008) Variation of nutrients in response to the highly dynamic suspended particulate matter in the Yangtze river plume. Cont Shelf Res 28:2393-2403.

Golterman HL (1996) Fractionation of sediment phosphate with chelating compounds: a simplification and comparison with other methods. Hydrobiologia 335:87-95.

Hans WP (2006) Assessing and managing nutrient-enhanced eutrophication in estuarine and coastal waters: Interactive effects of human and climatic perturbations. Ecol Eng 26:40-54.

Hasnain G, Levine BF, Gunapala, S, Chand N (1990) Large photoconductive gain in quantum well infrared photodetectors. IEEE 57(6):608-610.

Helen PJ, Colin N, Paul JA (2006) Withers. Sewage-effluent phosphorus: A greater risk to river eutrophication than agricultural phosphorus? Sci Total Environ 360:246-253.

Huang WZ, Schoenau JJ (1998) Fluxes of water-soluble nitrogen and phosphorus in the forest floor and surface mineral soil of a boreal aspen stand. Geoderma 81(3-4):251-264.

Hieltjes AHM, Lijklma L (1980) Fractionation of inorganic phosphates in calcareous sediments. J Envir Qual 9:405-407.

Javed I, Hu RG, Feng ML, Lin S, Saadatullah M, Ibrahim MA (2010) Microbial biomass, and dissolved organic carbon and nitrogen strongly affect soil respiration in different land uses: A case study at Three Gorges Reservoir Area, South China. Agric Ecosyst Environ 137:294-307.

Kristina LF, Adina P, Margaret LD (2005) Phosphorus distribution in sinking oceanic particulate matter. Mar Chem 97:307-333.

Lack TJ (1971) Quantitative studies on the phytoplankton of the rivers Thames and Kennet at Reading. Freshwater Biol 1(2):213-224.

Li FQ, Ye L, Liu RQ, Cao M, Cai QH (2008) Dynamics of main nutrient input to Xiangxi Bay of the Three-Gorges Reservoir. Acta Ecologica Sinica 28(5):2073-2079.

Li JX, Du B, Sun YS (2005) Effect of hydrodynamics on eutrophication. Water Resources and Hydropower Engineering 5:15-18.

Li MT, Xu KQ, Masataka W, Chen ZY (2007) Long-term variations in dissolved silicate, nitrogen, and phosphorus flux from the Yangtze River into the East China Sea and impacts on estuarine ecosystem. Estuar Coast Shelf Sci 71:3-12.

Li Y, Cao WZ, Su CX, Hong HS (2011) Nutrient sources and composition of recent algal blooms and eutrophication in the northern Jiulong River, Southeast China. Mar Pollut Bull 63:249-254.

Li ZP, Zhang TL, Chen BY (2006) Changes in Organic Carbon and Nutrient Contents of Highly Productive Paddy Soils in Yujiang County of Jiangxi Province, China and their Environmental Application. Agric Sci China 5(7):522-529.

Lu XX, Li SY, He M, Z Y, Bei RT, Li L, Alan DZ (2011) Seasonal changes of nutrient fluxes in the Upper Yangtze river basin: An example of the Longchuanjiang River, China. J Hydrol 405:344-351.

Lund E (1970) On the nomenclature of the pneumococcal types. IJSEM 20(3): 321-323.

Marina P, Donald FC (2007) Diatom metrics for monitoring eutrophication in rivers of the United States. Ecol Indic 7:48-70.

Meng QJ, Feng QY, Wu QQ, Meng L, Cao ZY (2009) Distribution characteristics of nitrogen and phosphorus in mining induced subsidence wetland in Panbei coal mine, China. Procedia Earth & Planetary Science 1:1237-1241.

Michael RP, Martin TA (1997) Seasonal variability in phosphorus speciation and deposition in a calcareous, eutrophic lake. Marine Geol 139:47-59.

Ogrinc N, Faganeli J (2006) Phosphorus regeneration and burial in near-shore marine sediments (the Gulf of Trieste, northern Adriatic Sea). Estuar Coast Shelf Sci 67:579-588.

Pastres R, Ciavatta S, Cossarini G, Solidoro C (2005) The seasonal distribution of dissolved inorganic nitrogen and phosphorous in the lagoon of Venice: A numerical analysis. Environment Int 31:1031-1039.

Paulo CFCG, Paul JW, Ian DM (2004) Seawater induced release and transformation of organic and inorganic phosphorus from river sediments. Water Res 38:688-692.

Ruban V, López-Sánchez JF, Pardo P, Rauret G, Muntau H, Quevauviller P (2001) Harmonized protocol and certified reference material for the determination of extractable contents of phosphorus in freshwater sediments: A synthesis of recent works. Fresen J Anal Chem 370:224-228.

Schindler DW (1978) Factors Regulating Phytoplankton Production and Standing Crop in the World's Freshwaters. Limnol Oceanogr 23(3):478-486.

Selles F, Kochhann RA, Denardin JE, Zentner RP, Faganello A (1997) Distribution of phosphorus fractions in a Brazilian Oxisol under different tillage systems. Soil Tillage Res 44:23-34.

Soyupak S, Mukhallalati L, Yemien D, Bayar A, Yurteri C (1997) Evaluation of eutrophication control strategies for the Keban Dam reservoir. Ecol Model 97:9-110.

Spitzy A, Ittekkot V (1991) Dissolved and particulate organic matter in rivers. In: Mantoura RFC, Martin JM, Wollast R (Eds.), Ocean Margin Processes in Global Change. John Wiley & Sons Ltd. pp. 5-17.

Sundby B, Gobeil C, Silverberg N, Mucci A (1992) The Phosphorus Cycle in Coastal Marine Sediments. Limnol Oceanogr 37(6):1129-1145.

Sutula M, Thomas SB, Brent AM (2004) Effect of Seasonal Sediment Storage in the Lower Mississippi River on the Flux of Reactive Particulate Phosphorus to the Gulf of Mexico. Limnol Oceanogr 49(6):2223-2235.

Swale EMF (1969) Phytoplankton in Two English Rivers. British Ecological Society 57(1):1-23.

Uselman SM, Robert GQ, Thomas RB (2000) Effects of increased atmospheric CO_2, temperature, and soil N availability on root exudation of dissolved organic carbon by an N-fixing tree (*Robinia pseudoacacia* L). Plant Soil 222(1-2):191-202.

Villares R, Carballeira A (2003) Seasonal variation in the concentrations of nutrients in two green macroalgae and nutrient levels in sediments in the Rıas Baixas (NW Spain). Estuar Coast Shelf Sci 58:887-900.

Vink S, Chambers RM, Smith SV (1997) Distribution of phosphorus in sediments from Tomales Bay, California. Marine Geol 139:157-179

Wang BD (2006) Cultural eutrophication in the Yangtze river plume: History and perspective. Estuar Coast Shelf Sci 69:471-477.

Wang P, He MC, Lin CY, Men B, Liu RM, Quan XC, Yang ZF (2009) Phosphorus distribution in the estuarine sediments of the Daliao River, China. Estuar Coast Shelf Sci 84:246-252.

Wang Y, Shen ZY, Niu JF, Liu RM (2009) Adsorption of phosphorus on sediments from the Three-Gorges Reservoir (China) and the relation with sediment compositions. J Hazard Mater 162:92-98.

Wu F, Qing H, Wan G (2001) Regeneration of n, p and si near the sediment water interface of lakes from southwestern China plateau. Wat Res 5(35):1334-1337

Xiang SL, Zhou WB (2011) Phosphorus forms and distribution in the sediments of Poyang Lake, China. Int J Sediment Res 26:230-238.

Xu CY, Gong LB, Jiang T, Chen DL, Singh VP (2006) Analysis of spatial distribution and temporal trend of reference evapotranspiration and pan evaporation in Changjiang (Yangtze River) catchment. J Hydrol 327(1-2): 81-93.

Yang HJ, Shen ZM, Zhu SH, Wang WH (2007) Vertical and temporal distribution of nitrogen and phosphorus and relationship with their influencing factors in aquatic-terrestrial ecotone: a case study in Taihu Lake, China. J Environ Sci 19: 689-695.

Yan TM, Yang LZ, Campbell CD (2003) Microbial biomass and metabolic quotient of soils under different land use in the Three Gorges Reservoir area. Geoderma 115:129-138.

Yao QZ, Yu ZG, Chen HT, Liu PX, Mi TZ (2009) Phosphorus transport and speciation in the Changjiang (Yangtze River) system. Applied Geochem 24: 2186-2194.

Young K, Morse GK, Scrimshaw MD, Kinniburgh JH, MacLeodd CL, Lester JN (1999) The relation between phosphorus and eutrophication in the Thames catchment, UK. Sci Total Environ 228:157-183.

Zeng H, Song LR, Yu ZG, Chen HT (2006) Distribution of phytoplankton in the Three-Gorges Reservoir during rainy and dry seasons. Sci Total Environ 367: 999-1009.

Zhong CH, Xing ZG, Zhao WQ, Wang DR, Deng CG, Li YJ, Xing M (2005) Eutrophication investigation and assessment of the Daning River after water storage in the Three Gorges Reservoir. Chinese Journal of Geochemistry 24(2): 149-154.

3

Hydrodynamic Effects

3.1 Overview

Eutrophication of the water body is a topic of widespread interest and the factors that cause it are mainly to do with nutrients, dissolved oxygen, water temperature, illumination and sediment (Wetzel and Limnology, 2001). Furthermore, hydrodynamics also play a crucial role in eutrophication. This consists of flux, flow velocity and water level. Hydrodynamics affect the water body mainly by wind drift and waves, which drive the sediment in the water body to move. Some kinds of substances in the water body then began to mix, dissolve, deposite, suspend, adsorb, entrain and cohere (Xu et al., 2009). In this condition, substances and sediments in the water body will change, which ultimately leads to water quality variety (Liebhold et al., 2004). When the flow velocity is fast, it is hard for eutrophication to occur even if the level of nutrients are high enough to trigger it, because some algae could be washed downstream by the flow before their growth has reached its peak (Zeng et al., 2006). Then the conditions for growth are destroyed and will not result in eutrophication. However, in other slow-flowing water bodies like lakes, reservoirs, estuaries, bays, inland seas, the flow velocity is slow and the water body is changing slowly (Smith, 1935). This condition slows down the spread of the nutrients and aggravates accumulation of the nutrients especially nitrogen and phosphorus, which offer fundamental nutrients for the growth and reproduction of algae. Furthermore, it can also offer appropriate hydraulic conditions for algae survival like green or blue green algae in water bodies which get used to a slow-flowing water body (Oliver and Ganf, 2000). Both suitable circumstances and a slow-flowing regime easily contribute to eutrophication of a still or slow moving water system. Most regions of China have a warm temperate continental monsoon climate and the wind force has big effect on water flow. In the season of relatively moderate wind, the current will move from bottom to top by wind force. During this process, nutrients in sediments will enter the water body in the water flow, which results in an increase in nutrients (Wagner, 2000). That is what an internal release of nutrients means. In this condition, algae will grow extensively, if other conditions are appropriate, and then eutrophication happens (Nadalal and Bogardi, 1995). In some shallow lakes and reservoirs the hydrodynamic effect causes sediment suspension and disturbs the sediment after large waves or a strong storm, which leads to nitrogen and phosphorus release (Sehelske, 2009; Liggett and Cunge, 1975; Schelske and Hecky, 2009). It can also cause the release of algae cells from sediment which then results in algal bloom (Sverdrup, 1953). Hydrodynamic conditions also play an essential role in migration and diffusion of pollutants. Streams which flow through a shallow lake produce many different patterns of water circulation by the action of wind power in the lake. The pattern and strength of the circulation will have large effect on migration and diffusion of pollutants in the lake along with the wind direction and change in the wind force. Furthermore, hydrodynamic conditions are closely related to the growth rate of algae itself. The content of algae will rise when the flow becomes slow under normal conditions (Buribrd et al., 2007). Generally, eutrophication will not happen in natural rivers which have characteristics of steep gradients and high flow velocity which make algae growth difficult (Escart and Aubrey, 1995). However, the condition is the reverse in a lake or reservoir and nutrients like nitrogen and phosphorus accumulate because of slow water flow, which offers large nourishment to algae (Mitrovic et al., 2003). Then, algae will grow quickly under proper climatic conditions, which make water turbidity rise and transparency decrease and algae bloom occur (Wu et al., 2009). The water body is heavily polluted under these conditions.

The construction of large-scale water conservancy projects makes humans efficiently manage and make full use of the water source. This brings many profits. A dam stores upstream water after construction, which meets the demands of flood control and provides irrigation and hydroelectric power and an improvement in river shipping conditions. It brings huge economic efficiency as described above. However, a dam changes the natural runoff of a river and causes the hydraulics to change. Upstream water is transformed into still water from a torrent. The change in the flow regime causes sediment of upstream of the dam to accumulate. The water and electricity demands change with the daily variation and seasonal dynamics, which lead to long or short-period alteration respectively for flood discharge. Then it causes discontinuity in natural rivers and the diversity of the river morphology declines. The whole ecological system in the river is destroyed under these conditions. The Three Gorges Reservoir (TGR) is located on the mid-downstream Yangtze River, China. Despite its benefits in terms of power generation and flood control and for shipping, the TGR has attracted attention for its potential impact on ecosystems and socioeconomic stability. The construction of the TGR has changed the natural runoff of the Yangtze River, which will inevitably bring many ecological problems such as eutrophication in the reservoir, silting in the reservoir, a fall in the water level, a water flow temperature change, supersaturated or insufficient dissolved gas, channel erosion downstream of the dam, changes to anadromous fish reproduction, geological hazards and so on.

The total scheduling mode of the Three Gorges reservoir has two kinds of operation modes including the one for the flood season and the other one after it. In the flood season it operates along with the water level dropping to 145 meters in order to meet the requirements of flood control. In this case the reservoir has a lower water level and a larger flow and bigger flow velocity and its resultant effect is the same as in natural conditions. Water in the region of the reservoir and downstream are rather lightly polluted. After the flood season (from November to April) the reservoir operates at a high level of 175 m. It creates one reservoir which is 660 km long and is 1.1 km wide and 90 m deep and with 39.3 billion m^3 capacity. The reservoir waters changed fundamentally after construction, the water surface was wider, the water was deeper and the flow velocity was slower. The sediment accumulated in large amounts, the flow velocity at the tail of the mainstream reservoir was less than 0.5 m/s and at the head and core of the reservoir was about 0.5 m/s too (Zeng et al., 2006). Therefore, in this case algae bloom would most likely occur if other conditions were appropriate. The reservoir had a smaller quantity of water after the flood season and the station could afford more cycling operations which could enlarge the amount of the downstream flow with the water level of the inner and downstream reservoir region fluctuating and with the water flow vibrating, which increase the diffusing capacity of pollutants and moderate the pollution of the reservoir region and its downstream waters. Cycling operations at the station were the main measure of ecological changes at the Three Gorges reservoir. When the reservoir was operating at a high level of 175 m after the flood season, the station operated in variable load mode. The two factors changed the flow regime of the natural water-course and triggered environmental hazards (Shen et al., 2008). To sum up, the main ecological problems revealed the following three aspects. Firstly, the water level became higher and foreign flow into the reservoir and the downstream flow became slower and the flow velocity slowed. In particular, in a branch stream and some local regions in the reservoir the water even appeared stagnant. In this case, the water decreased the dilution and diffusion capacity of pollutants, which led to water quality degradation. Secondly, the cycling operation makes the downstream flow fluctuate significantly during one single day and causes the water level between the Three Gorges Dam and the Gezhou Dam to frequently fluctuate and dramatically change in flow velocity, which is not profitable for marine shipping and creates shipping problems. Lastly, the reservoir water level rises and the flow velocity of the tail region falls. So sediment is deposited in the tail reservoir and causes water to flow back upstream. Eutrophication occurred in innumerable river embayments on a large or small scale after the Three Gorges reservoir was constructed. The main reason was that the area of cross section increased and the flow velocity decreased enormously along with sediment deposits which cause water quality changes and enlarged water transparency, which is convenient for the photosynthesis of algae. In particular, minor tributaries in the reservoir were affected by the water level of the mainstream. The flow velocity of the backwater area was also slow (almost below 0.01 m/s) and sediments were deposited. Eutrophication had happened at the end of the branch-stream and became serious.

Many researchers regarded TP and TN as the potential causes of eutrophication in the mainstream of the reservoir, but the flow conditions in the branch-stream and reservoir variations caused by water storage were the main inciting factors (Platt, 2001). The flow velocity of the reservoir was rather quick and the water body was deeper and the optical energy loss was greater in the natural watercourse before construction, which do not promote algae growth. However, the flow conditions changed a lot. For example, the flow velocity slowed down quickly, which led to the deposit of large sediments and the increase in water transparency and light transmittance. These factors promoted the photosynthesis of algae which accelerate algae growth. Eutrophication was rather easy to achieve in this situation. The effects described above were the direct inducible factors of eutrophication. This viewpoint was accepted many researchers. Furthermore, they pointed out that both water storage and a slow current were conducive to water bloom. Moreover, the difference in water characteristics between the mainstream and branch-stream of the reservoir were obvious. Water level adjustments cause the water to interflow between the mainstream and the branch-stream. In turn the exchange will further change the reservoir's physicochemical characteristics. It affects the water bloom occurring in the branch-stream of the reservoir. Therefore, research into the relationship between flow velocity and algae growth was the theoretical basis for avoiding water eutrophication. Many factors were closely connected with eutrophication, apart from flow velocity, such as area, volume, depth, shoreline coefficients, incoming runoff supply coefficients, residence period, water level variation, outbound runoff and so on. Besides, the importance of the ecological hydrological mechanism of lakes on eutrophication has been largely accepted by many researchers. For example, Leman et al. indicated that hydrological conditions could regulate effectively diatom blooms. They also successfully changed phytoplankton community structure by controlling hydrologic conditions. Jone et al. showed that hydrologic conditions and nutrients could affect phytoplankton community structure and abundance. Recently, the effect of hydrodynamic conditions on algae growth has aroused research interest. Mclntire had proved flow had significantly impacted on predominant species of algae through a laboratory modeling experiment. Steinman and Mclntire (1986) revealed that low flow rates were more beneficial to eutrophication. Escart and Aubrey (1995) had pointed out flow rates had some effect on the algae growth rate. In China, most research focused on the effect of wind-drive current on the algae growth rate. Li et al. (2004) indicated wind-driven current could drive sediment to the bottom of the lake, which leads to the fast release of nutrient from the sediment to the water body. Yan et al. (2008) reported algae had the best growth rate under conditions of 90 r/min disturbance velocity through a study of the hydrodynamic force of water circulation on algae growth. In total, domestic and overseas studies mostly focused on the effect of wind-driven current and horizontal flow velocity on the algae growth rate. But fewer studies were about the impact of slow vertical velocity caused by falling water levels triggered by leakage at the base of the hydroelectric reservoir. In actual circumstances, hydrodynamic conditions in the common water body also consist of the flow rate of the longitudinal section, turbulence intensity and surface. fluctuations. In addition, research with the focal point on how hydrodynamic conditions impact on the algae growth rate did not have a common standard because of different experimental conditions, various experimental methods and circumstances.

TN and TP contents were rather high in the Three Gorges reservoir and its main rivers and had exceeded the internationally recognized criterion of eutrophication

(Yan et al., 2008). When comparing the period before and after the impounding of the Three Gorges reservoir, algae now grew in more favourable situations because of a falling flow rate, though its nutriment content fell a bit, which meant eutrophication happened more easily. Algae chlorophyll content of the reservoir bay was much higher than in the mainstream and the first signs of eutrophication were more obvious. Many researches had showed that eutrophication caused mainly by diatoms and dinoflagellates had happed many times in some bays, especially in the spring. Water storage in the Three Gorges Reservoir reached 135m in June 2003. However, before long the Daning river water quality became abnormal and some of its backward areas had been heavily eutrophicated. There are seven rivers backwater areas in which eutrophication had happened to different degrees in 2004. There is a tendency to increase the nutriment ranking of the Daning River compared to the report of 2003. There are 27 reservoirs.

Through the above description, it is known that eutrophication is closely related to nutrients, dissolved oxygen and hydrodynamic processes. Therefore, this chapter will focus on the impact of hydrodynamic conditions on the distribution of sediment, including sediments vertical concentration distribution and the vertical particle size distribution. Moreover, the conversion processes of nitrogen and phosphorus in sediments and water is also discussed.

3.2 Hydrodynamics Processes

There are three types of sediment transport in a river including imminent transport, slight transport and full transport according to the theory, which respectively correspond to different pollutant concentrations (Adi and Benny, 2005). So the types of sediment release are obviously different. When the flow rate is slow, the mud surface remains static (Ding and Susan, 2002; Wang et al., 2005). Then the surface begins to change along with a rising flow rate. Firstly, the force of the whole sludge becomes large because of the flow speeding up (Tsai et al., 2003). Then a small zone of diluents above the sludge begins to stop and this reveals imminent movement. Then the sediment reaches its second period of transport with the flow rate rising (Zhao et al., 2003). Sludge in a small area of some riverbeds is scattered, which drives the sediment in other areas and makes the water body obviously turbid (Xia and Pignatello, 2001; Piatt et al., 1996). When the flow rate speeds up to 60–70 m/s, the river bottom fluctuates fiercely. This is the third period of sediment transport. It can be seen that the sludge is stirred up constantly and the smooth riverbed is largely destroyed. If those parts that are destroyed, the sediment is torn to pieces. Soon the water becomes completely turpid. A improved syntonic turbulence-simulation device (ISTSD) (Fig. 3.1) was used to simulate the natural aquatic environment, and perform hydrodynamic experiments to estimate the effect of hydrodynamic condition on resuspended particles.

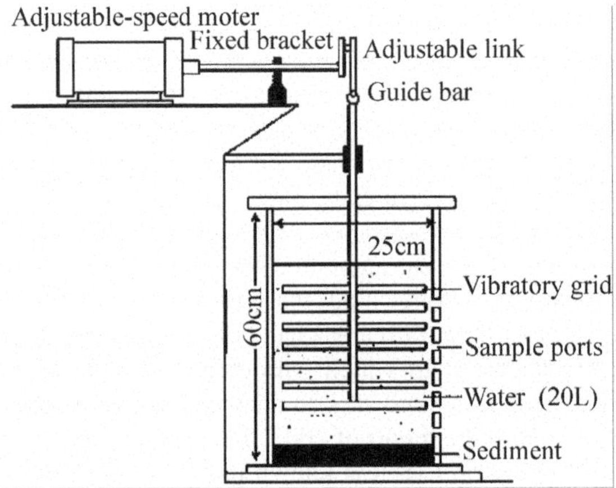

Fig. 3.1 Schematic diagram of improved turbulence-simulation device

When the sediment was in the condition of imminent transport, the sediment release rate was on the rise along with the flow rate rise. TN and TP concentration and the release rate rise a small amount because the sediment is only disturbed slightly and the amount of suspensed sediment is small. So only the nutriments in interstitial water are released and cannot affect the water characteristics efficiently. When the flow rate increases more and the sediments are at the second stage, TN and TP concentration and release rate increases clearly compared with the previous stage. Now some sediment begins to be stirred up and there are small amounts of suspended sediment in the upper water, which carry the nutriments into the water body. Meanwhile, the interstitial water in the bottom sediment is released in large amounts. All the factors above lead to the increase in TN and TP concentration. When the sediment is at the third stage, TN and TP concentration and release rate show a sudden increase and all the nutriments in the sediment are released into the water body. TN and TP concentration is three times more and seven times more than the initial amount, respectively. The release rate is ten to twenty times that of the initial rate. Similarly, Reddy revealed that the nutriment content in the upper water caused by suspension, including suspension and diffusion, increased ten times more than that caused only by diffusion when he studied Apoka lake. Sondeaard et al. reported that nutriment content caused by dynamic suspension increased twenty to thirty times more than the initial rate when he did a field survey in Areso Lake in Denmark. This research has proved that hydrodynamic forces play a great role in the endogenous N and P cycles in lakes.

The Three Gorges reservoir has had obvious effects on the ecological water environment (Guo et al., 2009). Its hydrological regime has changed and had an impact on the ecological systems after the first tentative water storage in June 2003. Some branch-stream pollution became heavier and eutrophication happened in some of its branch-stream backwaters to different degrees. Much research has shown hydrodynamic conditions were the main factors for eutrophication and for algae blooms to appear (Hsiao et al., 2001). Zen et al. analyzed phytoplankton's seasonal changes and the relationship to nutriments and hydrological conditions in the Three Gorges reservoir using field surveys and laboratory simulation experiments. When Huang and Walter (1998) studied the correlation between the flow rate and algae growth in the Daning River according to the dynamic theory of rivers, they also made use of Saint-Venant's principle by simulating its hydrodynamic conditions and set up a model for the relationship between flow rate and algae growth (Gerard and Orjan, 2006; Serrano et al., 2005). The study of hydrodynamic characteristics of the branch stream of the Three Gorges reservoir reduces to a one dimensional model after its impounding. Then its effects on phytoplankton and eutrophication were analyzed and the model for eutrophication and warning of water blooms was constructed based on this (Myrna et al., 2005). However, in the actual situation, the water surface in the branch stream was almost horizontal and the water flow was weak under the force of gravity (Karickhoff et al., 1979). There would be a remarkable lag between the detecting result and the actual one if we use the one dimensional model to analyze and warn about eutrophication and the water bloom problem. Therefore, it should be made clear what measures concerning hydrodynamic conditions were taken in the Three Gorges reservoir before doing research on the eutrophication and water bloom problem. In total, the experimental impounding period in the Three Gorges reservoir was mainly divided into the following three stages. Firstly, the water level target was 135 m when the reservoir was filled for the first time in June 2003. The second water level target was 156 m in October 2006 and the third water level target was 175 m in September 2009 (Hu and Cai, 2006). The Three Gorges reservoir operated mainly by impounding clear water and releasing the muddy flow. Specifically speaking, it enlarged the flow rate through depressing the water level of reservoir during the period of high sediment in the reservoir waters. The reservoir regulates the amount of sediment by releasing water downstream. It raises the water level and restores water in the reservoir when the inflow of water has only a little amount of sediment.

The Daning River is the one of the main branch streams and lies at the core of the Three Gorges reservoir. Its drainage area is 4045 km^2. It belongs to the subtropical zone enjoying a humid monsoon climate. It has many typical characteristics such as four distinct seasons; mild weather in winter and earlier spring, a hot summer with a drought, rain in autumn, heavy humidity, heavy clouds and slight wind force. The average temperature is 16.6 ℃ and average rainfall is 1124 mm (Zhen et al., 2009). Then the dynamic processes need to be clarified taking the Daning River as an example. There were three stages in the experimental impounding period of the Three Gorges reservoir, described as above. The water level and the runoff shown as Figs. 3.2a and 3.2b were the water levels before the Three Gorges reservoir construction and its inflow, respectively. The hydrologic regime of the Daning River showed a big change after impounding in June 2003 and it was affected by the upstream inflow and the mainstream of the lake. Fig. 3.2a shows that the reservoir operated at a high water level during the high water level operation period and maintained a low water level during the flood season. The water level fluctuated slightly in the two periods. The water level decreased mainly in the leakage phase and the increase focused on the storage period. Fig. 3.2b shows the inflow into the reservoir had a smaller annual trend varying from 10015.63–

16226.92 m^3/s, but had a larger annual variation with its maximum always occurring in the flood season, with a daily flow being 26347.06 m^3/s, followed by a storage period and leakage period. The minimum occurred in the high water level operation period and the daily inflow was almost 6003.03 m^3/s. According to the Three Gorges reservoir project and water level before the reservoir construction and inflow before flowing into the reservoir, the water level should undergo the following five complete cycles: high water level operation period from November to April, water leakage period from May to July, limiting flood period from July to September, storage period in September to November and then the high water level operation period. In total, the flow rate of the Daning River was quite slow and its average value was measured in centimeters. Some monitoring point even had a flow rate of almost 0 m/s. The Daning River flows in a southeast direction. Meantime, the absolute value of the mainstream flow rate was much larger than the lateral velocity. Therefore, only the mainstream flow rate characteristics were addressed in this book. Fig. 3.3 is the longitudinal profile of the Daning River in the typical months. During the investigating period in 2010, the Daning River was affected by the twin influences of the upstream inflow and mainstream pounding and there was always the phenomenon of the waterflow moving in different directions not flowing in a one-dimensional direction. January and March 2010 represented the high water level operation period. The mainstream water flow into the reservoir was mainly in the form of bottom undercurrents which are below 30 m. The range of water intrusion was 10 km away from the river mouth in January and increased more in March to affect the middle and upper Daning River. May 2010 represented the leakage period and the amount of undercurrent flow from the upper river to the mouth increased. The mainstream water flowed into the reservoir by a surface layer which was between 0 and 30 m of flow backwards. August 2010 represented the limiting flood period, then the mainstream water flowed into the reservoir by the middle layer which was between 5 and 40 m of flow backwards. September 2010 represented the storage period. The mainstream also flowed in backwards but its undercurrent range was large enough to affect the upper. In total, there was always an undercurrent flowing from the upper river to the estuary during the exploration period. All estuaries exhibited the same phenomenon of the mainstream flowing backwards into the reservoir. However, the backward submergence depth and range were influenced by climate, like water temperature and air temperature and hydrographic factors like upper runoff and water level variation.

3.2.1 Parameters of Turbulence-Simulation Device

Parameters of the turbulence-simulation device were calibrated according to the vertical distribution data of sediment concentration under different hydrodynamics conditions. The relationship between the turbulent diffusion coefficient and vibration frequency of the turbulence-simulation device was also calculated. Therefore, the relationship between hydrodynamic conditions and vibration frequency

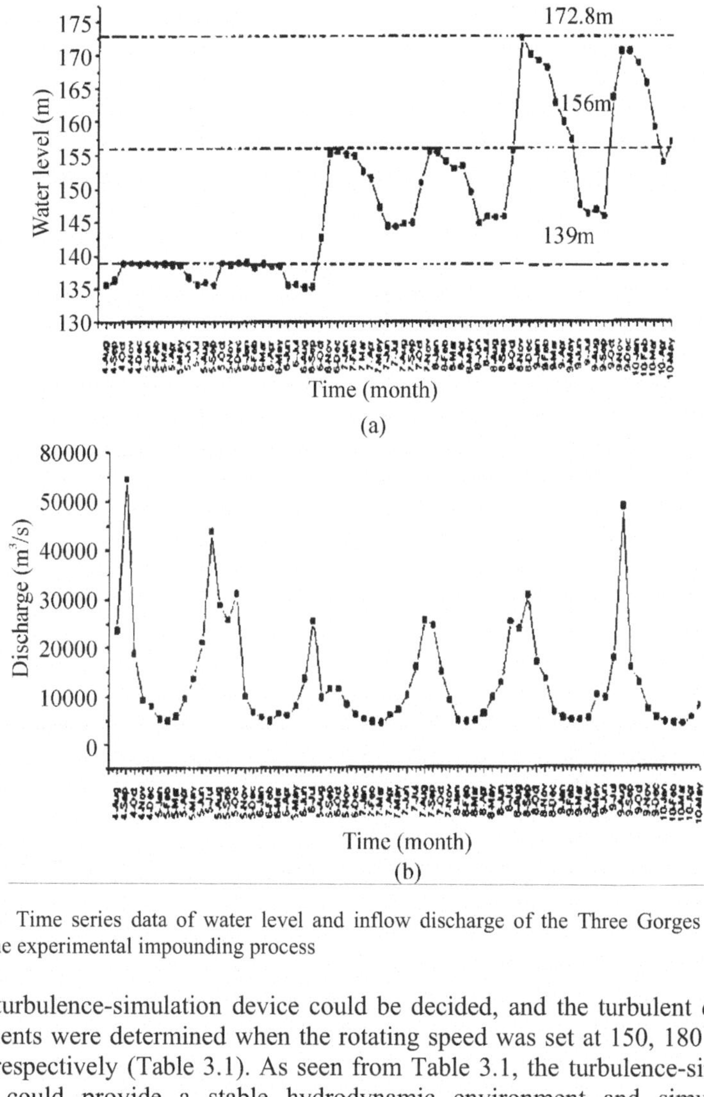

Fig. 3.2 Time series data of water level and inflow discharge of the Three Gorges Reservoir during the experimental impounding process

of the turbulence-simulation device could be decided, and the turbulent diffusion coefficients were determined when the rotating speed was set at 150, 180 and 280 r/min, respectively (Table 3.1). As seen from Table 3.1, the turbulence-simulation device could provide a stable hydrodynamic environment and simulate the hydrodynamic conditions efficiently. Here, the hydrodynamic condition was expressed as rotating speed.

Table 3.1 The relationship between the rotating speed of turbulence-simulation device and turbulence intensity

Rotating speed(r/min)	Sediment concentration(kg/m^3)	Relation	ε
150	1.004	ε=9.71 F	24.28
180	2.878	ε=9.49 F	28.47
280	6.797	ε=9.88 F	46.11

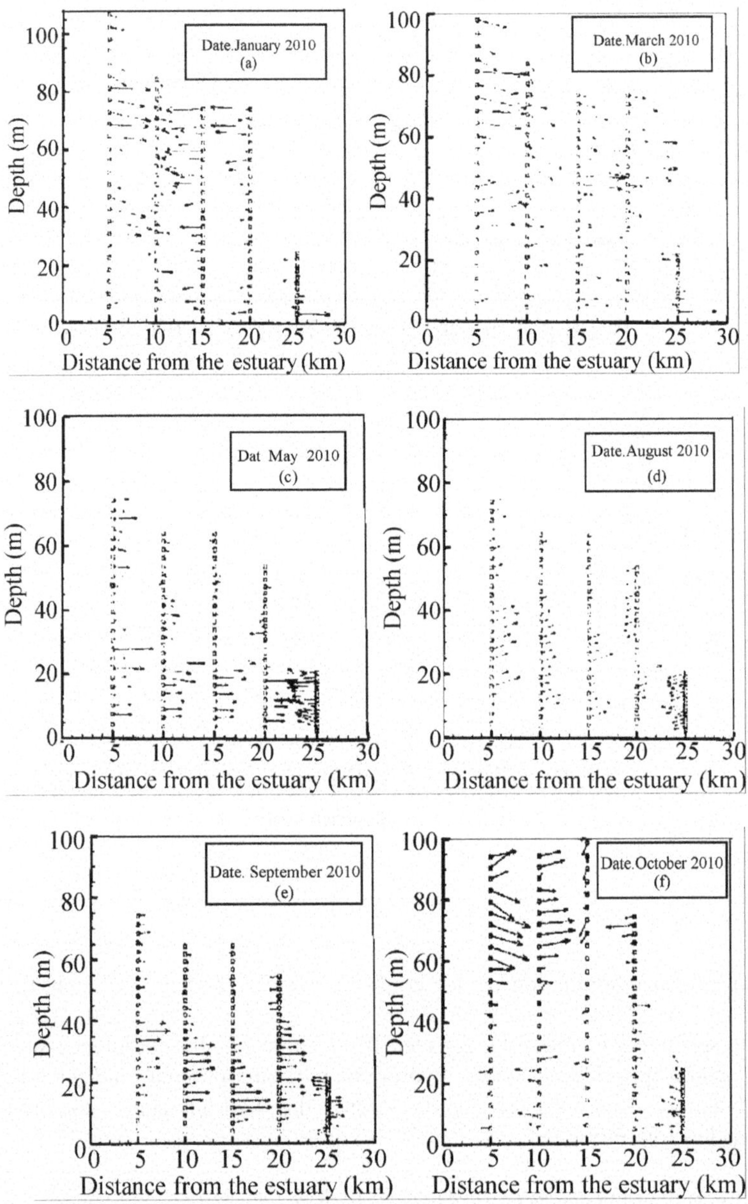

Fig. 3.3 Dynamics characteristics of velocity at longitudinal profile in the Daning River

3.2.2 *Vertical Distribution of Sediment Concentration Under Different Hydrodynamic Conditions*

Suspensions at different depths of the turbulence-simulation device were sampled for analyzing the sediment concentration under different hydrodynamics conditions. Different hydrodynamics conditions were expressed at different rotating speeds of the turbulence-simulation device, which was set at 150, 180 and 280 r/min, respectively.

Comparing the sediment concentration in the turbulence-simulation device under different hydrodynamics conditions (Table 3.2; Wang et al., 2009) with those of the real water environment of The Yangtze River (measured values obtained from some hydrological stations, Table 3.3), the results indicated that the turbulence-simulation device could efficiently simulate the real hydrodynamics and sediment conditions.

Table 3.2 The sediment concentration at different sampling depths under different hydrodynamic conditions

Sampling depth (cm)	Sediment concentration (kg/m^3)		
	150 r/min	180 r/min	280 r/min
30	0.90	2.57	6.11
25	0.80	2.70	6.65
20	1.10	2.94	6.70
15	1.02	3.05	6.72
10	1.08	3.20	7.41
5	1.12	3.41	7.19

Moreover, the vertical distribution curve of sediment concentration at different rotating speeds was also drawn. The sampling depth varied from 5 to 30 cm, which was relative to the bottom of the turbulence-simulation device. As seen from Figs. 3.4a, 3.4b, 3.4c, at the three designed rotating speeds, the sediment concentration increased with the decrease in sampling depth, but the vertical distribution variation of the sediment concentration was not obvious, which was similar to the hydrological conditions of the Wuhan section of the Yangtze River. The gap in water potential at the Wuhan section is low, resulting in the sediment concentration being relatively stable and the vertical distribution of sediment concentration varying very little. Furthermore, the relationship curve of the sediment concentration and rotating speed was also obtained. The turbulent diffusion coefficient of the turbulence-simulation device increased with the increase in rotating speed, and the total sediment concentration showed a tendency to increase (Fig. 3.5).

Table 3.3 The sediment concentration and median particle diameter of samples from some hydrological stations on the Yangtze River

Hydrological stations	Pingshan	Yichang	Hankou	Datong
Mean annual sediment concentration (kg/m^3)	1.760	1.140	0.573	0.486
Maximum annual sediment concentration (kg/m^3)	2.890	1.650	0.772	0.697
Minimum annual sediment concentration (kg/m^3)	1.060	0.610	0.264	0.280
Mean annual median particle diameter (mm)	0.031	0.022	0.018	0.017
Maximum annual median particle diameter (mm)	0.063	0.050	0.039	0.042
Minimum annual median particle diameter (mm)	0.013	0.006	0.007	0.008

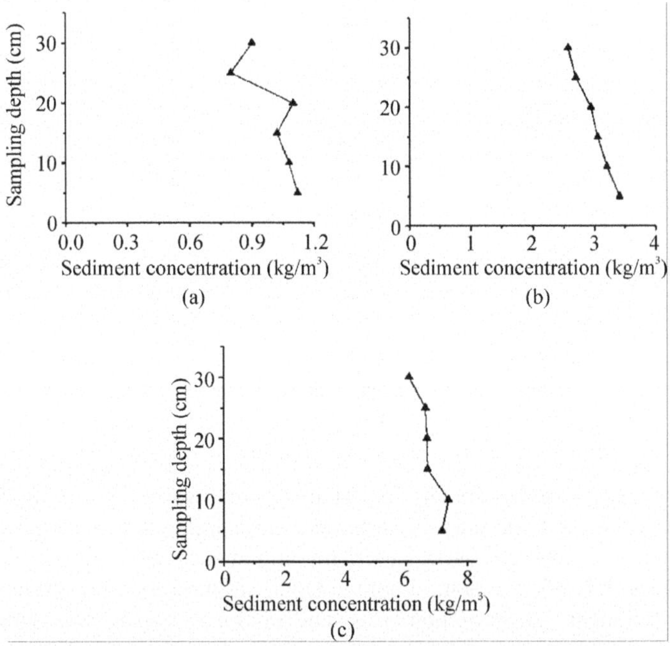

Fig. 3.4 The vertical distribution curves of sediment concentration at 150 r/min (a), 180 r/min (b) and 280 r/min (c), respectively

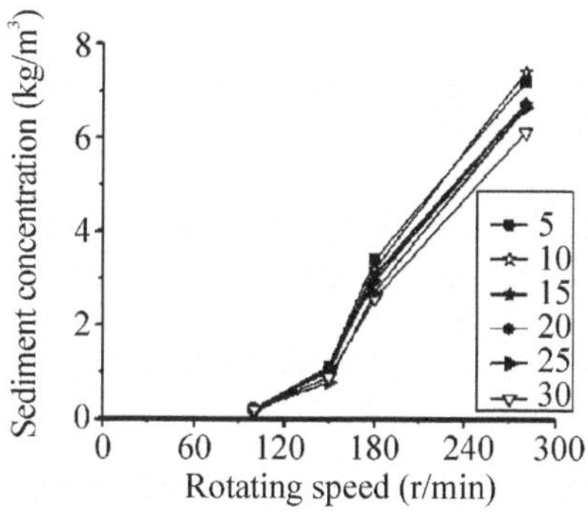

Fig. 3.5 The relationship curve of the sediment concentration and rotating speed of the turbulence-simulation device

3.2.3 Vertical Distribution of Sediment Particle Size Under Different Hydrodynamic Conditions

The particle size of suspended sediment under different hydrodynamic conditions was analyzed. The results demonstrated that the median particle diameter of suspended sediment at the fixed sampling depth increased with the increase in the rotating speed of the turbulence-simulation device. This may be mainly attributed to the increase in the turbulent diffusion coefficient, which could aggravate the strength of motion of water and resuspend the large particles on the bottom, resulting in more large-size sediment particles and the median particle diameter of suspended sediment increasing. The median particle diameter of suspended sediment at different rotating speeds was shown in Table 3.4 (Wang et al., 2009), which was similar to that in the real water environment of the Yangtze (Table 3.3), indicating that the turbulence-simulation device could simulate the real hydrodynamic conditions efficiently again.

Table 3.4 The median particle diameter of suspended sediment under different hydrodynamics conditions

Rotating speed (r/min)	Median particle diameter (μm)					
	30 cm	25 cm	20 cm	15 cm	10 cm	5 cm
150	5.10	5.13	5.10	5.11	5.12	5.25
180	8.99	9.03	9.43	9.42	9.52	9.52
280	16.99	17.15	17.20	17.35	17.28	17.64

The particle size of suspended sediment under different hydrodynamics conditions has been divided into four ranks, including Φ<0.025 mm, 0.025 mm<Φ<0.063 mm, 0.063 mm<Φ<0.125 mm and 0.125 mm<Φ<0.2 mm. The distribution of the sediment particle size at different sampling depths was concluded in Fig. 3.6 (Wang et al., 2010). As seen from the figure, at the same rotating speed the distribution of the sediment particle size at different sampling depths was similar; however, it varied significantly when the rotating speed was different. The ratio of large-size particles increased and that of small-size particles obviously decreased and thus the median particle diameter of suspended sediment increased gradually. It can be inferred that the hydrodynamic conditions played a very important role in the sediment concentration and the distribution of sediment particle size. The transportation of substances, such as nitrogen, phosphorus and some organic pollutants, was influenced by the sediment concentration and the distribution of the sediment particle size. Therefore, the hydrodynamic conditions were the important influencing factor on the transportation of nitrogen and phosphorus in water and sediment.

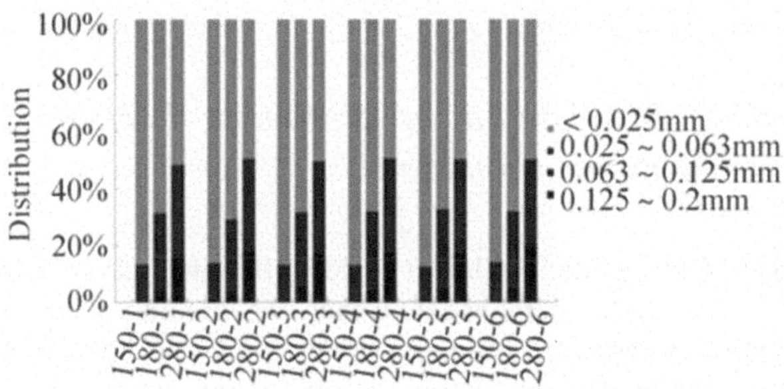

Fig. 3.6 The distribution of sediment particle size under different hydrodynamics conditions

3.3 Transformation of Nitrogen

Nitrogen is one of the most important macro nutrients for aquatic organisms. However, discharge of nitrogen into receiving waters can lead to a significant impact on water quality which, with high levels of phosphorus, can cause excessive growth of phytoplankton and eutrophication of water bodies (Lefebvre et al., 2011). The rates of nitrogen uptake (by plants) and the effects on aquatic organisms are different from one nitrogen species to another (McCutcheon, 1987; Xia et al., 2008). Nitrification is an aerobic biological process in which ammonia (NH_3) is oxidized to nitrite (NO_2^-) or (NO_3^-) via autotrophic ammonium oxidation, a two-step process carried out by two categories of chemolithotrophic microorganisms: ammonia-oxidizing bacteria and nitrite-oxidizing bacteria. Denitrification is an anoxic process in which NO_3^- and NO_2^- are reduced to N_2 gas via nitric oxide (NO) and nitrous oxide (N_2O) in a process requiring an organic carbon source. Nitrification, which was considered to be a surface-based process in water and wastewater systems (Rittmann and McCarty, 1981; 1978; Gantzer et al., 1988), is a fundamental process of the nitrogen cycle in aquatic systems. Previous studies indicate that nitrifying bacteria tend to grow by attaching to the surface of sediment particles (Dunnette and Avedovech, 1983; Belser, 1979). This is especially true in shallow streams (Kusuda et al., 1994; Lau, 1990). Increasing evidence has demonstrated the important role of sediment-based nitrification in rivers and lakes (Butturini et al., 2000; Blackburn et al., 1994). For example, Pauer and Auer (Pauer and Auer, 2000) noted that the nitrification rate was rapid in the sediments, whereas lack of nitrification was observed in the water column of a hyper-eutrophic lake and the adjoining river system. (Gribsholt et al., 2005) measured the total system and water column nitrification of a tidal freshwater marsh and showed the dominance of particle/sediment related nitrification.

Xia et al. (2008) indicated that nitrogen pollution is one of the most critical problems for surface water quality in China, due mainly to high concentrations of ammonium nitrogen (NH_4^+-N). Some reaches of some rivers cannot be used as drinking water sources because their NH_4^+-N concentrations are higher than 1.0 mg/L (71 μmol/L) with the water quality worse than grade III on the Chinese water quality grade scale, in which grade I is best and grade V is worst (Xia et al., 2004). For China, nitrification is particularly important as a natural process that converts ammonia from wastewater and leads eventually to removal of nitrogen from surface waters. Therefore, the goal of their research on the Yellow River was to study the effect of suspended sediment concentration on the nitrification rate. To study the effect of the state of the sediment on nitrification, the collected water and SPS were used as media for nitrification experiments; the indigenous bacteria in the samples remained active. Three stirring regimes, i.e. no stirring, intermittent stirring (stirring for 12 h and without stirring for 12 h), and continuous stirring, were carried out to study the state of the sediment on the nitrification rate. The initial NH_4^+-N concentration was 5.0 mg/L (357 μmol/L) and the SPS concentration was 5.0 g/L in each water-sediment system and the flasks were then covered by eight layers of gauze to exclude external bacteria, then incubated in the dark at 20 °C with a magnetic stirrer to ensure oxygen saturation and to promote the growth of indigenous nitrifying microorganisms as well as to encourage interactions between water and the SPS. Aliquots were withdrawn for the determination of concentrations of nitrogen species. The nitrification rate was calculated based on NO_3-N concentrations in the water phase and/or total NH_4^+-N concentrations in both water and SPS phases. Each flask and the gauze were sterilized before the experiment. Each experimental set was conducted in triplicate with a set of controls. The control experiments were carried out by adding 0.5% mercuric chloride. To study the effect of the SPS-water interface on the nitrification processes, the collected water and SPS were sterilized by autoclaving for 30 min at 120 °C, and used as media for cultivation; the ammonia and nitrite oxidizing bacteria isolated from the water and SPS samples were introduced into the system, with both the ammonia-oxidizing and nitrite-oxidizing bacteria populations being 105 cell/mL and the initial NH_4^+-N concentration being 786 μmol/L in each water system. The cultivation experiments were then carried out as described in the above section. The nitrification rate was higher with the presence of suspended-sediment than deposited-sediment. The sequence of the nitrification rate was "no stirring" <"intermittent stirring"<"continuous stirring"; the zero-order kinetic rate constants of nitrification (based on NO_3-N variations) were 12, 33 and 60 μmol N/(L·day) during the first two days, respectively. Since stirring suspended the sediment and led to an increase in the exposed surface area of the SPs, the above results support the view that the increase in the exposed surface area of the SPs would stimulate the nitrification processes. This was consistent with the results obtained by (Abril et al., 2000) where the sediment and fluid mud resuspension/settling resulted in high nitrification and denitrification rates

3.3.1 Ammonia Adsorption in Sediments

We introduce the influence of suspended particle concentration on ammonia adsorption.

As NH_4^+ is an electrically positive ion species, it is very easy to be adsorbed into the SPs which are negatively charged. The dependence of ammonia adsorption on different concentrations of SPs was evaluated. Surface water used in this study was filtered through a 0.45 μm microvoid filter film to remove suspended particulate matter and plankton. A series of conical flasks (100 mL) were added with a given mass of sediment samples and 30 mL of NH_4Cl solution in concentrations of 0.5, 1.0, 2.0, 5.0, 10.0 and 20.0 mg/L. Flasks were then sealed, and shaken at 150 r/min in a constant temperature shaker (T = 20 °C). Aliquots of 1.0 mL were collected at an interval of 10 min until equilibrium was obtained. The suspensions were centrifuged at 4000 r/min for 20 min and the NH_4Cl concentration in the supernatant was measured with photometric analysis.

The preliminary experiments revealed that the adsorption of NH_4^+ onto SPs was very fast and approached equilibrium within 30 min, so that in this test the measure of supernatant NH_4^+ was conducted after 1 h of adsorption reaction. Fig. 3.7 shows the adsorption of NH_4^+ over a range of suspended particle concentrations varying from 1.0 to 10 g/L (Wang et al., 2010). The variations in SPs used here are typically occurring in water systems of the Three Gorges Reservoir (Yangtze River Water Resources Commission, 2000). At four suspended particle concentrations, the ammonia adsorption linearly increased with the increase in initial ammonia concentrations (R^2 = 0.98–0.99), but the increasing rate and the amount of ammonia adsorbed per unit weight of suspended particles were greater at lower concentrations of suspended particles. Under the experimental conditions of different suspended particle concentrations and initial ammonia concentrations, the ammonia adsorption was quite effective. Specifically, the transfer rate of ammonia from solution to sediments reached 99% at 2.0 mg/L of NH_4^+-N irrespective of the SPs concentrations. Even with ammonia loading increasing to 20 mg/L of NH_4^+-N, the adsorption efficiency still approached 92% for four SPs concentrations, resulting in residual ammonia concentrations of 1.24–1.56 mg/L of NH_4^+-N in supernatant solution. The high adsorption efficiency of ammonia under the experimental conditions may also explain the linear relationship between the ammonia adsorption and the initial ammonia concentrations.

According to the above observation that ammonia loading of up to 20 mg/L of NH_4^+-N was quickly and effectively fixed by the suspended particles with typical concentrations present in the TGR, it can be expected that once the external ammonia at a given concentration enters the water body of the TGR, it should be firstly adsorbed by the SPs. As thus, the ammonia adsorption by SPs may have a significant effect on biological transformations of nitrogen cycling, such as nitrification/denitrification. This research result is consistent with the observation previously made that the nitrification in the Ems estuary and Onondaga Lake mainly took place in SPs and bed sediment. Moreover, the ammonia fixed by the SPs will become an important internal nitrogen pollution source as a consequence of sedimentation of suspended particles in the TGR.

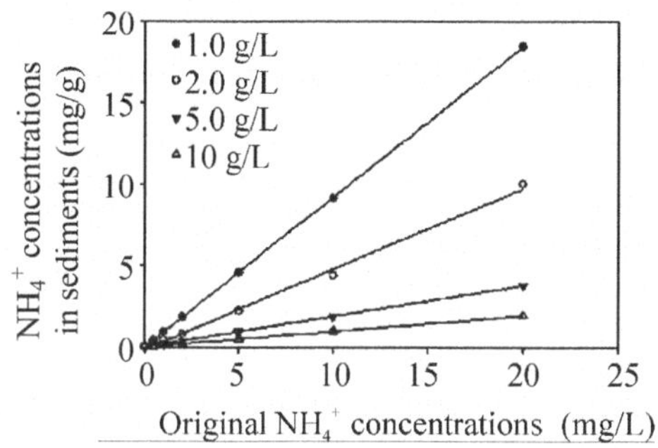

Fig. 3.7 Effect of suspended particle content on the ammonia adsorption (T=25 °C)

3.3.2 Ammonia Adsorption with Different Particle Sizes and Organic Matter Contents in the SPs

The size distribution of suspended particles broadly fluctuated as a result of the change in the hydrodynamic conditions in various water bodies. The organic matter content of suspended particulate originating from various sources is different. The adsorption of ammonia in suspended particles of different sizes and different organic matter content was investigated, as shown in Fig. 3.8 (Wang et al., 2010) and Fig. 3.9. For five suspended particle sizes and four different organic matter contents, the ammonia adsorption linearly increased with the increase in initial ammonia concentration (R^2 = 0.99). Neither the particle size nor organic matter content of SPs showed a significant effect on ammonia adsorption under the experimental conditions. This suggests that the variation in particle size and organic matter content will not change the adsorption capacity of ammonia under typical environmental conditions in the TGR. However, the two factors may influence the biological transformation of nitrogen occurring in SPs, and they are investigated in the following chapters.

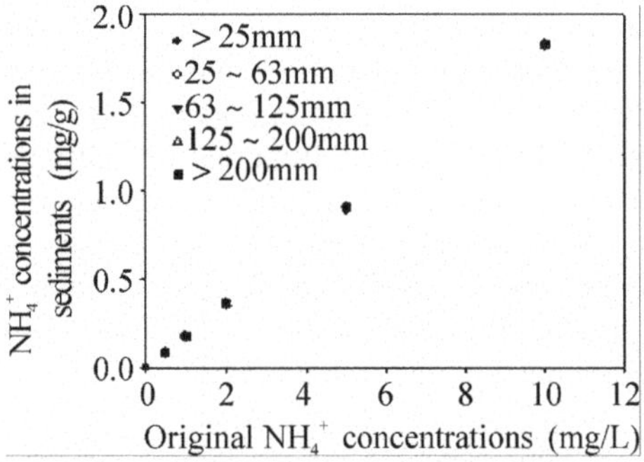

Fig. 3.8 Ammonia adsorption on suspended particles with different sizes (T=25 °C, sediment concentration=2 g/L)

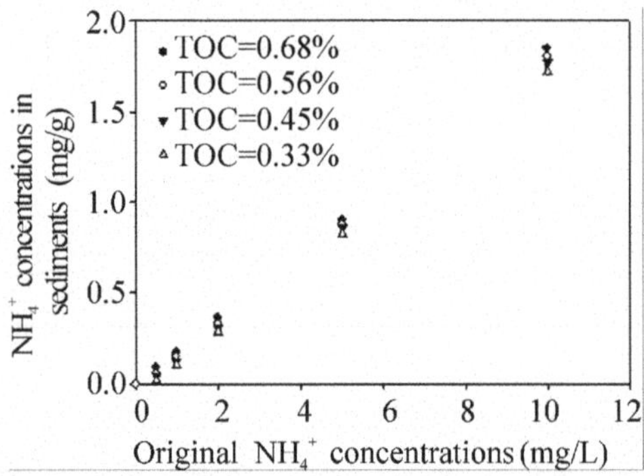

Fig. 3.9 Effect of organic matter contained in suspended particles in the ammonia adsorption (T=25 °C, sediment concentration=2 g/L)

Laboratory microcosm experiments were used to estimate the influence of suspended particles on the adsorption of NH_4^+. The adsorption of NH_4^+ was quite effective under the conditions of different suspended particle concentrations and initial ammonia concentrations, which typically occurred in the water systems of the Three Gorges Reservoir. Even with ammonia loading increasing to 20 mg/L of NH_4^+-N, the adsorption efficiency still approached 92% for four SPs concentrations (1.0, 2.0, 5.0 and 10 g/L). For four studied suspended particle sizes and four experimental organic matter contents, the reflected influences on the adsorption of ammonia were not obvious under the experimental conditions. The quick and effective adsorption of ammonia under the experimental conditions implied that the ammonia pollution inflowing into the water body would firstly be fixed on the suspended particles and consequently be loaded into the sediments.

3.4 Transformation of Phosphorus

Phosphorus in lakes and reservoirs comes mainly from weathering and erosion products in the river basin and soil and rock, atmospheric sedimentation, industrial and agricultural production and domestic sewage. It could be divided into two catagories: endogenous phosphorus and exogenous phosphorus according to the different sources. Exogenous phosphorus comes mainly from industrial and agricultural production, human waste and soil erosion. Endogenous phosphorus comes mainly from the sediment release from lakes and biodegration. In the process of eutrophication in lakes, a large amount of nutriments pours into the lakes, which leads to the accumulation of superfluous nutriments in lake sediment and biological bodies and is the main source of endogenous phosphorus. Many studies reported eutrophication also had to do with the endogenous pollution from sediments in lakes except for exogenous nutriment inflow. In the process of managing lake eutrophication, it was common that endogenous water characteristics still could not improve after exogenous phosphorus release was controlled because of endogenous phosphorus release. The main reason for this was that endogenous release of the dynamic and static phosphorus keeps the lake always at the eutrophication level. Total phosphorus in the water body exists mainly in the form of soluble phosphate and particle suspended solids. Soluble phosphate mainly includes all forms of orthophosphoric acid and a small amount of organophosphorus. But phosphorus in particle suspended solids is rather complex and maybe itself is one form of insoluble phosphate such as all kinds of algae, bacteria, microbes, dead plants and animals (Zhang et al., 2001). It was also likely to make all kinds of clay or mineral particles the main framework which combined together through humus and a metal oxide hydrate adhesion bridging force, which adsorbed particular matter including soluble phosphate (Karim and Husain, 2010).

The source of the force among the material in the sediments has two aspects based on physical and chemical theory: one is a kind of chemical force whose action range is close to the solid surface, such as a covalent bond, hydrophobic bond, hydrogen bond, steric hindrance and directional effect. The other is a kind of force whose action range is further away like electrostatic attraction and Van der Waals force. These forces were determined by the main framework of the sediment and the physical and chemical characteristics of the adsorbate. The material in the suspended matter was desorbed or the suspended matter flocculated and cohered together when the outer force was big enough to overcome any force source above (Du et al., 2000), which finally led to the transformation of the physical characteristics of the suspended matter. In the static environment there is no flow rate, only gravity and buoyancy that affect suspended matter and those two forces were equivalent, that is they could not have had the function of working on the suspended matter. It means that material in the suspended matter would be suspended anywhere in the water body in the natural circumstances and have no transformation. However, the flow produces a force on the suspended matter in the water and drives the suspended matter with the water flow. This force is commonly called shear stress. Meantime, suspended particles were also affected by water resistance and a striking force caused by a different rate of motion, except for gravity and flocculation, when moving with the water flow (Liang et al., 2011). The material was affected simultaneously by all of these forces working in nonequivalent conditions which could work on them and overcome their attractive force. The result was obviously that simple adsorbate separated from the suspended material mainframe became simple ions then soluble in the water. These suspended materials combined with each other and were then transformed into the new one. Therefore, suspended material with a core organism had a smaller density, especially algae, and flocculated on the water surface on account of its buoyancy. Nevertheless, suspended material with a core of clay had a much higher density and finally overcome the buoyancy and was deposited. Certainly, the results above were obtained in conditions where the flow rate was rather slow. When the flow rate continued to increase, it not only affected the suspended material but also affected the sediment in the water.

Imminent transport means flows don not work on the particles in the sediment but only work on the material in the water. In the state of slight transport, the flow rate begins to work on the sediment and the sediment can be seen. The result is that the phosphorus in the sediment can enter the water body along with suspended material. Furthermore, it can be seen that the water and water body become turbid in conditions of full transport. In these circumstances, it is believed that the flow rate cannot drive the transport of sediment when the flow rate is slow, meaning that sediment has an imminent transport status. But some flocculant deposits carry away parts of the phosphorus, which leads to the first transport period. When the phosphatic mass concentration in overlying water is quite high under the conditions of stirring, the sediment is the accumulation point not the source. As the flow rate continues to rise, the sediment moves into the slight transport period. At this time, the particles released from the sediment are larger in number than the deposited particles caused by flocculation, so the phosphorus in particles in the inflow is greater than the phosphorus carried away. The mass concentration of total phosphorus increases and the second transport stage is reached. The later phase of the second period is the period before full transport. The role of hydraulics enlarges the density of the particles in the water body and makes the deposited particles rise in number. This retards the increase in particles in the water to some extent. It means that the increase is reduced in the later phase of the second period. When the flow rate is rather large, the sediment is transported entirely and the water body becomes turbid. At this time, the total phosphorus concentration in the water body is close to that in the sediment, so there is soon a spurt in the growth period. Soluble phosphorus is the main component in the water body. Its mass concentration was low in the static environment and its relative proportion almost doubled and increased to a maximum of almost 87% at a velocity of 12.5 cm/s. Afterwards, its relative proportions become smaller as the velocity continues to increase. The soluble phosphorus concentration does not decrease but increases as the total phosphorus concentration decreases under the conditions of slow flow, which shows that the current cannot drive the sediment but it can release the soluble phosphorus from the sediment. The relative proportions of soluble phosphorus mass concentration obviously decrease when the flow rate is between 12.5 cm/s and 30 cm/s. However, the decrease is over a small range as the flow rate increases. The main reason is that sediment moves in large amounts under the conditions of a strong current.

References

Abril G, Riou SA, Etcheber H, Frankignoule M, De Wit R, Middleburg JJ (2000) Transient, tidal scale, nitrogen transformations in an estuarine turbidity maximum-fluid mud system (the Gironde, South-West France). Estuar Coast Shelf Sci 50:703-715.

Adi O, Benny C (2005) Sorption-desorption behavior of polycyclic aromatic hydrocarbons in upstream and downstream river sediments. Chemosphere 16:19-29.

Belser LW (1979) Population ecology of nitrifying bacteria. Annu Rev Microbiol 33:309-333.

Blackburn TH, Blackburn ND, Jensen K (1994) Risgaard-Petersen N. Simulation model of the coupling between nitrification and denitrification in a freshwater sediment. Appl Environ Microbiol 60:3089-3095.

Buribrd M, Jolmson S, Cook A (2007) Correlations between watershed and reservoirs characteristics and algal blooms in subtropical reservoirs. Water Res 41:4105-4114.

Butturini A, Battin TJ, Sabater F (2000) Nitrification in stream sediment biofilms: the role of ammonium concentration and DOC quality. Water Res 34:629-639.

Ding XL, Susan MH (2002). Adsorption and desorption of proteins and polyamino acids by clay minerals and marine sediments. Mar Chem 77:225-237.

Du WS, Fang D, Yang JR (2000) Study on incipient velocity of sediment. J Hydraulic Eng 10:51-56.

Dunnette DA, Avedovech RM (1983) Effect of an industrial ammonia discharge on the dissolved oxygen regime of the Willamette River, Oregon. Water Res 29:1121-1127.

Escart HJ, Aubrey DJ (1995) Flow structure and dispersion within algal mats: Estuar Coast Shelf Sci 40:451-472.

Gantzer CJ, Rittmann BE, Herricks EE (1988) Mass transport to streambed biofilms. Water Res 22:709-722.

Gerard C, Orjan G (2006) Effects of added PAHs and precipitated humic acid coating on phenanthrene sorption to environmental black carbon. Environ Pollut 141:526-531.

Gribsholt B, Boschker HTS, Struyf E, Andersson M, Tramper A, De Brabandere L,van Damme S, Brion N, Meire P, Dehairs F (2005) Nitrogen processing in a tidal freshwater marsh: A whole ecosystem 15N labeling study. Limnol Oceanogr 50:1945-1959.

Guo W, Xia Z, Wang Y (2009) Ecological operation goals for Three Gorges Reservoir. Adv Water Sci 20:554-559.

Hsiao LH, Whei M, Lee G (2001) Enhanced naphthalene solubility in the presence of sodium dodecyl sulfate: effect of critical micelle concentration. Chemosphere,

44:963-972.

Hu ZY, Cai QH (2006) Preliminary Report on Aquatic Ecosystem Dynamics of the Three Gorges Reservoir Before and After Important. Acta Hydrobiologica Sinica 30(1):1-6.

Huang WL, Walter JW (1998) A distributed reactivity model for sorption by soils and sediments. II. Slow concentration-dependent sorption rates. Environ Sci Technol 32:3549-3555.

Karickhoff SW, Brown DS, Scott TA (1979) Sorption of Hydrophobic Pollution Nature Sediments. Water Res 13:241-248.

Karim Z, Husain Q (2010) Application of fly ash adsorbed peroxidase for the removal of bisphenol A in batch process and continuous reactor: Assessment of genotoxicity of its product. Food Chem Toxicol 48:3385-3390.

Kusuda T, Futawatari T, Oishi K (1994) Simulation of nitrification and denitrification processes in a tidal river. J Water Sci Technol 30:43-52.

Lau YL (1990) Modelling the consumption of dissolved contaminants by biofilm periphyton in open channel flow. Water Res 24:1269-1274.

Lefebvre A, Guiselin N, Barbet F, et al. (2011) Long-term hydrological and phytoplankton monitoring (1992-2007) of three potentially eutrophic systems in the Eastern English Channel and the Southern Bight of the North Sea. ICES J Mar Sci 68:2029-2043.

Li YP, Feng Y, Chen KS (2004) Study on the starting principles of sediment by water force in Taihu lake. Adv Water Sci 15:770-774.

Liang RF, Han L, Doraiswamy D, et al. (2011) The Rheology of Aramid Platelet Suspensions. Polym Eng Sci 51:1933-1941.

Liebhold A, Koenig W, Bjornstad O (2004) Spatial synchrony in population dynamics. Auu Rev Ecol Evol Syst 35:467-490.

Liggett JA, Cunge JA (1975) Numerieal methods of solution of the unsteady flow equations. Unsteady Flow in Open Channels, Water Resources Publications.

McCutcheon S (1987) Laboratory and in-stream nitrification rates for selected streams. J Environ Eng 113:628-646.

Mitrovic SM, Oliver RL, Rees C (2003) Critical flow velocities for the growth and dominance of anabaena circincilis in some turbid freshwater rivers. Freshwater Biol 48:164-174.

Myrna JS, Benny C, Ashish PD, et al. (2005) Comparison of polycyclic aromatic hydrocarbon distributions and sedimentary organic matter characteristics in contaminated, coastal sediments from Pensacola Bay, Florida. Mar Environ Res 59:139-163.

Nadalal KDW, Bogardi JJ (1995) Operating of a reservoir for quality control using inflows and outflows. J Water Sci Technol 31:273-280.

Oliver RL, Ganf GG (2000) The Ecology of Cyanobacteria. The Netherlands: Kluwer Academic Publishers.

Pauer JJ, Auer MT (2000) Nitrification in the water column and sediment of a hypereutrophic lake and adjoining river system. Water Res 34:1247-1254.

Piatt JJ, Backhus DA, Capel PD, et al. (1996) Temperature-dependent sorption of naphthalene, phenanthrene and pyrene to low organic carbon aquifer sediments. Environ Sci Technol 30: 751-760.

Platt J (2001) Reservoir use estimation modeling with water lever fluctuation. U.S. Department of the Interior, Bureau of Reclamation.

Rittmann BE, McCarty PL (1978) Variable-order model of bacterial film kinetics. J Environ Eng 104:889-900.

Rittmann BE, McCarty PL (1981) Substrate flux into biofilm of any thickness. J Environ Eng 107:831-849.

Schelske D, Hecky R (2009) Eutrophication: More nitrogen data needed. J Science 324:721-722.

Serrano S, Garrido F, Campbell CG, et al. (2005) Competitive sorption of cadmium and lead in acid soils of Central Spain. Geodema 124:91-104.

Shen Z, Qian H, Hong Y (2008) Parameter uncertainty analysis of the non-point source pollution in the Daning River watershed of the Three Gorges Reservoir Region, China. Sci Total Environ 405:195-205.

Smith H (1935) Synchronous flashing of fireflies. Sci 82:151-152.

Steinman AD, McIntire CD (1986) Effects of current velocity and light energy on the structure of perphyton assem blags in laboratory streams. J Phycol 22:352-361.

Sverdrup H (1953) On conditions for the vernal blooming of phytoplankton. ICES J Mar Sci 18:287-295.

Tsai WT, Lai CW, Hsien KJ (2003) The effects of pH and salinity on kinetics of paraquat sorption onto activated clay. Colloid Surf A 224:99-105.

Wagner M (2000) Effect of hydrological patterns of tributaries on biotic processes in lowland reservoir-consequences for restoration. Ecol Eng 16:79-90.

Wang HY, Shen ZY, Guo XJ, Niu JF, Kang B (2010) Ammonia adsorption and nitritation in sediments resourced from Three Gorges Reservoir. China Environ Geol 60:1653-1660.

Wang LL, Shen ZY, Wang HY, Niu JF, Lian GX, Yang ZF (2009) Distribution characteristics of phenanthrene in the water, suspended particles and sediments from Yangtze River under hydrodynamic conditions. J Hazard Mater 165(1-3): 441-446.

Wang ZY, Zhai ZC, Wang LS (2005) QSPR modeling of adsorption coefficient KOC of alkyl(1-phenylsufonyl) cycloalkane-carboxylates on soil and sediments using MLSER model and abinitio. J Mol Struct 732:79-85.

Wetzel GR, Limnology R (2001) Lake and River Ecosystems. London: Academic Press.

Wu XD, Kong FX, Chen YW (2009) Horizontal distribution and transport process of bloom-forming microcystis in a large shallow lake. Limnol 2:1-10.

Xia GS, Pignatello JJ (2001) Detailed sorption isotherms of polar and apolar compounds in a high-organic soil. Environ Sci Technol 35:84-94.

Xia XH, Yang ZF, Huang GH, Maqsood I (2004) Integrated water quantity and quality evaluation of the Yellow River. Water Int 29:423-431.

Xia XH, Yang ZF, Zhang XQ (2008) Effect of suspended-sediment concentration on nitrification in river water: importance of suspended sediment-water interface. Environ Sci Techol 43:3681-3687.

Xia X, Li S, Shen Z (2008) Effect of Nitrification on Nitrogen Flux across Sediment-Water Interface. J Water Environ Res 80:2175-2182.

Xu Y, Cai Q, Shao M (2009) Seasonal dynamics of suspended solids in a giant subtropical reservoir (China) in relation to internal processes and hydrological features. Quatern Int 208:138-144.

Yan R, Feng Y, Zhao W (2008) Influence of circumfluent type waters hydrodynamics on growth of algae. China Environ Sci 28:813-817 (in Chinese).

Yan Q, Yu Y, Feng W (2008) Plankton community composition in the Three Gorges Reservoir Region revealed by PCR-DGGE and its relationships with environmental factors. J Environ Sci 20:732-738 (in Chinese).

Zeng H, Song L, Yu Z (2006) Distribution of phytoplankton in the Three Gorges Reservoir during rainy and dry seasons. Sci Total Environ 367:999-1009.

Zhang L, Fan CX, Qin BQ (2001) Phosphorus Release and Absorption of Surficial Sediments in Taihu Lake under Simulative Disturbing Conditions. J Lake Sci 13:35-42.

Zhao XK, Yang GP, Gao XC (2003) Studies on the sorption behavior of nitrobenzene on marine sediments. Chemosphere 52:917-925 (in Chinese).

Zhen B, Cao C, Zhang J, Huang M, Chen Z (2009) Analysis of Algal Blooms in Daning River of Three Gorges Reservoir. Environ Sci 30:3218-3226.

4 Biological Effects

4.1 Overview

Water eutrophication admits superfluous nutrient substances (nitrogen and phosphorus), which lead to the abnormal growth of algae and other aquatic life, changes in water diaphaneity and dissolved oxygen. All the factors above accelerate water substance ageing and affect the aquatic ecosystems and water body functions.

There are various factors which lead to eutrophication, including physics, chemistry and biological agents. In all of these factors, various living organisms play an important part in the eutrophication and the transmission of substances and energy.

In the natural water body, the ecosystem is composed of animate and non-animate organisms. The organisms can be divided into producers, consumers and disintegrators. The producers mostly include algae and aquatic plants which can absorb the nutrients in water to photosynthesize organic matter and translate solar power into chemical energy. This process can manufacture organic matter and supply oxygen and food for other aquatic life. A producer would assimilate quantitative nutritive elements in the process of photosynthesis and growth, which contributes to water purification. A consumer, like water fowls, can control the growth of floating grass by expending organic matter and energy. Therefore, all the organisms in the water body are indispensable for the ecosystem. In the water body algae and other organisms which are influenced by nitrogen and phosphorus, especially in the pent-up water shell, will breed without intermission and spend a great quantity of dissolved oxygen in the water, whereas the death and decomposition of an algoid will cause the nitrogen and phosphorus to be released into the water body again and engender a vicious circle of algoids bred in the water body. Microorganisms play a great role in both the circulation of materials and the energy flow in the ecosystem. In other words, organic compounds can enter the recycling process with the help of different types of microbes.

Biological oxidation of organic compounds mainly relies on the role of heterotrophic microbes. These microbes feed on organic compounds and decompose them into simple compounds, which can offer the material and energy for bacterial growth and reproduction. Simple compounds which exist in the form of a true solution can enter the inner bacterial cells. However, complex organic compounds of high molecular weight can't be absorbed directly by bacteria and they must be hydrolyzed into simple compounds, which can be absorbed into the body for metabolism, using the external enzymes. Therefore, decomposition and assimilation of organic compounds in the water environment must involve the hydrolization and bio-oxidation process of cellular metabolism. All of these processes are related to the microbial species.

Microbial ecology is a branch of science which researches the relationships between the microorganism and the biotic and abiotic environment around it. The content of microbial ecology is mainly dealing with microbial species, their distribution in the natural environment, their laws of variation with the changes in environmental conditions, and so on.

Environmental microbes are the community composed of many microbial populations. There exist many relationships, such as symbiosis, mutualism, co-residency, competitive relationships and so on, among different populations, which play a big role in the material circulation and the energy conversion process. At the same time, microbes can be the indicators which are related to eutrophication (Lesniewska and Witak, 2011; Lodi et al., 2011), and have some effect on the degree of eutrophication (Du et al., 2011; Xu et al., 2011). Microbial structure and function is one of the research topics of microbial ecology. The research into the microbial community can be done at several different levels including microbial quantity, metabolic activity, community structure and metabolic function. The community structure, metabolic function and their relationships to the microbial population are the core of the research into the microbial community. Microbes in the water are closely associated with their regional environment and play a big role in the circulation of matter and the energy conversion process. Their quantity and population distribution are closely concerned with many factors like the type of water, organic content and microbial inhibitory effects. The study of the microbial community structure and metabolic activity can reveal the biological basis of the removal and transformation process of environmental pollutants, and can offer essential information on assessing and predicting environmental quality and safety. The study can also help to comprehend and evaluate the effect and mechanism of bioremediation and biological treatment technology. Meanwhile, it offers a theoretical direction on how to control and optimize the microbial community and strengthen its metabolic activity.

There mainly are two aspects to ecosystem research of lakes or reservoirs. One is research about the ecological structure of lakes or reservoirs. Its object is to make clear what and how many microbes exist in lakes or reservoirs. The other is research on the microbial function. It should figure out what role these microbes play in the circulation of matter and energy conversion processes of lakes or reservoirs.

Microbes play a great part in the ecological systems of lakes or reservoirs as the primary actor in l circulation of matter and energy conversion processes. However, it is difficult to study the microbial ecological system of lakes or reservoirs because of its special characteristics, such as small size, complexity of the changes, difficulties of observation, whereas this is easy in animals and plants (Gasol and Duarte, 2000). So far, many methods have been devised to investigate the microbial community of an ecosystem. One way of exploring microbial diversity is to isolate microbes in samples using a proper medium, and then explore the microbial community by strain identification. It is relatively reliable to analyze the microbial community by morphological observation and physiological and biochemical tests (Borsodi and Kurdi, 1998). But the experimental workload and cost are both tremendous and cultured microbe numbers only occupy 1% of the total number (Vigdis et al., 1990). Meanwhile, microbial morphology is limited and lacks obviously external features, so exploring the microbial community structure and function only by morphological observation and physiological and biochemical tests is far from enough. Recently, analysis of DNA diversity of microbes by genetic fingerprint technology has become highly valued with the application of molecular biological technologyl in microbial ecology (Yurkova et al., 2002). By this method, total microbial DNA can be directly extracted from samples and analyzed. Microbial genetic information including uncultured organisms can be obtained to the greatest possible degree when there is no need to isolate and culture strain. Therefore, microbial diversity of samples can be explored in full-scale aspects. This method can also offer reliable information about population. It is simple and feasible in detecting a specific group in the microbial community and assessing the diversity of isolated strains. However, there are many defects concerning molecular biology technology, for example information obtained would be large enough to block its analysis while exploring a complex microbial community. Some research showed that the sequencing results obtained in pure culture conditions were not in accord with those obtained by directly examining a sample.

In total, microbes play a great role in global biogeochemical cycles because of their wide distribution, diverse metabolic ability and efficient enzyme activity, and this activity might decide the potential productivity of a given ecological habitat, in large part. The changes in physical and chemical properties caused by human activity will impact the microbial community (including microbial species, distribution and richness) in the ecological habitat. So the species, distribution, richness and changes in functional microbes can directly reflect the features of the habitat, levels of certain matter and its rules of transformation. Therefore, microbes can work as potential bio-indicators to reflect the condition of the water because of some special characteristics such as small size, large specific surface area of a cell, diversity in physiological and biochemical functions, strong metabolic ability, ease in adapting to the new environment, sensitivity to environmental change and so on. In particular, the natural environmental conditions are quite complex and are in a constant state of variation, which lead to obvious differences in microbial structure and the ratio of different populations. Even though in the same environment, the composition of the microbial community and the ratio of various groups will change with the variation in environmental conditions. In other words, microorganisms are generally highly sensitive to the surrounding environments and are profoundly alerted by their perturbations. The water quality is strongly influenced by microbial community dynamics and ecosystem functions, such as organic matter content and nutrient recycling.

The term functional microbes (this book refers to functional bacteria) refers to a group of bacteria performing the same function. There are many kinds of functional bacteria groups, such as protein degradation bacteria, fat degradation bacteria, cellulose degradation bacteria, starch degradation bacteria, phosphate-solubilizing bacteria, nitrifying bacteria, denitrifying bacteria and so on. The functional bacteria play a crucial role in the ecological environment. On the one hand functional groups often present special growth conditions, on the other hand they perform special functions in the process of transformation of matter.

Water body eutrophication, in response to hyper-nutrification by N and P loads, is directly related to N and P cycling in aquatic environments. All major transformations of N and P in the environment are carried out exclusively by microbes including functional bacteria associated with N: nitrogen-fixing bacteria (NFB), ammonia-oxidizing bacteria (AOB), nitrite-oxidizing bacteria (NOB), ammonifying bacteria (AB), denitrifying bacteria (DNB), nitrate-reducing bacteria (NRB), and functional bacteria associated with P: inorganic phosphate-solubilizing bacteria (IPB) and organophosphate-solubilizing bacteria (OPB). It is predicted that the functional bacteria associated with N or P would be potential indicators in monitoring the eutrophic conditions.

The Three Gorges Project is one of the biggest water conservancy projects in the world and attracts the focus of the world. Hence, the eco-environmental problem of the Three Gorges Reservoir gets attention from all over the world. Different degrees of eutrophication appeared in some of the arterial sub-fluvial backwater areas, which become one of the biggest eco-environmental problems. A microorganism is an important ingredient of an aquatic ecosystem. It not only can decompose the organic matter in the water and sediments but can also act as the indicator to reflect the degree of pollution of the water body and the regenerative condition of the polluted water. As well known, eutrophication is mainly caused by two nutrient elements, nitrogen and phosphorus. The cycles of the two nutrient elements in the aquatic environment are mainly performed by the functional microorganisms. Due to the important role of microorganisms in aquatic ecosystem, we performed a series of researches on the microbial community in the backwater areas of the TGR. In this chapter, the characteristics of the microbial community in the TGR, especially the distributions of functional bacteria associated with N or P, were investigated. Furthermore, the effect of microbes on the water condition in the backwater areas of the Yangtze River in the TGR was explored.

4.2 Biological Zones

4.2.1 Sampling Locations and Properties

Surface water, bottom water and sediments in seven locations in the backwater areas of the Yangtze River in the TGR were sampled in October 2006, five at Daning River and the remaining two at the Xiaojiang River and Xiangxi River, the three main tributaries of the Yangtze River in the TGR. Surface water and bottom water (0–5 cm above the sampling sediments) were collected using sterile Niskin bottles. Sediments were collected with a Van Veen stainless steel grab sampler (Eijkelamp, Netherlands). Only the top 5 cm of sediments were used and stored at −20 °C prior to analysis. To count the total cells, parts of each sample were fixed with glutaraldehyde (final concentration 2%) immediately after collection.

The sampling locations and their physicochemical characteristics are listed in Table 4.1 (Wang et al., 2010). The pH values of surface water varied from 7.4 to 8.6, which was slightly higher than that of bottom water with the pH values ranging from 7.0 to 8.4. The temperature of surface water was between 19.5 and 20.5 °C, about 3 °C higher than that of bottom water. The DO in surface water, varying from 7.55 to 9.52 mg/L, was obviously higher than that in bottom water, varying from 5.15 to 5.63 mg/L. TN in surface water was obviously higher than that in bottom water, whereas TP showed no obvious difference between surface and bottom water.

Table 4.1 Sampling locations and physicochemical characteristics of surface water and bottom water

Position	Sampling location	Sample No.	pH	*T* (°C)	DO (mg/L)	TN (mg/L)	TP (mg/L)	Chl-a (mg/L)
Daning River	31°11.442' N 109°52.466' E	DN1-SW	7.6	20.5	7.55	1.05	0.06	1.03
		DN1-BW	7.4	16.5	5.20	0.73	0.05	
Daning River	31°08.642' N 109°53.774' E	DN2-SW	8.2	21.0	8.29	1.77	0.08	1.51
		DN2-BW	8.0	16.5	6.96	1.39	0.05	
Daning River	31°07.528' N 109°53.901' E	DN3-SW	8.4	21.5	9.25	2.03	0.09	12.14
		DN3-BW	8.2	0.05	0.92	6.90	17.0	6.66
Daning River	31°06.643' 109°53.465'	DN4-SW	8.2	20.0	8.05	3.81	0.09	ND
		DN4-BW	8.1	17.5	7.60	0.63	0.07	ND
Daning River	31°05.403' 109°53.590'	DN5-SW	8.4	19.8	9.52	2.07	3.43	0.80
		DN5-BW	8.3	16.0	8.67	0.72	2.35	12.14
Xiaojiang River	30°57.089'	XJ-SW	7.4	19.5	7.75	0.76	2.29	6.66
		XJ-BW	7.0	1	5.15	0.06	0.08	ND
Xiangxi River	30°58.146' N 110°45.639' E	XX-SW	8.6	7.5	7.81	0.14	0.06	ND
		XX-BW	8.4	19.5	7.63	0.11	0.10	

[a] SW: surface water; BW: bottom water; ND: not detected

Previous research indicated that the N element mainly came from exotic pollution and the P element originated from both exotic pollution and the release of sediments in the TGR (Wang, 2006; Zhang et al., 2007a). That could explain the difference in the distributions of N and P in water bodies. Chl-a concentrations in freshwater were relatively high, ranging from 0.80 to 12.14 mg/L, which were higher than those of eutrophication standards. The concentrations of Chl-a in the Xiaojiang River (XJ) and Xiangxi River (XX) locations, which were 6.66 and 12.14 mg/L respectively, both exceeded the super-eutrophication standards (Zhang et al., 2007b). It could be concluded that eutrophic conditions of XJ and XX were more severe than in the Daning River (DN). In total, according to the water quality index, it was inferred that the water eutrophic level of sampling places was XX>XJ>DN4>DN5>DN2>DN3>DN1, namely, Xiangxi River>Xiaojiang River> Daning River.

4.2.2 Culturable Bacteria Number on Different Nutrient Level Mediums

The total numbers of bacteria were determined using the acridine orange direct count (AODC) method as described by Coolen and Overmann (Coolen and Overmann, 2000). The numbers of cultivable bacteria were measured using the plate count technique, and the eight kinds of functional bacteria associated with N or P, namely AB, NFB, AOB, NOB, NRB, DNB, OPB and IPB, were determined by using both the Most-Probable-Number (MPN) method (AOB, NOB, NRB and DNB) and plate count technique (AB, NFB, OPB AND IPB). During plate count processes, each of the bacteria samples was diluted to different optimum concentration to acquire the most accurate colony estimation, which in our case was estimation from plates containing 30–300 colonies. The media and incubation conditions used for the enumeration of bacteria are listed in Table 4.2 (Wang et al., 2010). Variance in count data was normalized using log(x) transformation prior to analysis, where x equaled the average number of colony forming units (CFUs or MPN) per gram of dry sediments or (CFUs or MPN) per milliliter of freshwater.

The numbers of total bacteria (counted by the acridine orange direct count method, AODC) and culturable heterotrophic bacteria grown in various nutrient levels of the medium, including eutrophic-medium (beef extract peptone medium, BEP medium), rich-nutrient medium (PTYG medium), low-nutrient medium (20% PTYG medium) and oligo-nutrient medium (R2A medium), were counted (shown in Fig. 4.1 except R2A medium). Since no colony was cultured on R2A medium, its count result was not presented in Fig. 4.1 (Wang et al., 2010). The total bacteria were 8.12×10^6, 2.70×10^7, and 6.73×10^{10} CFUs per mL or per g dry weight for surface water, bottom water and sediments, respectively, indicating that the richness of bacteria was rather high in the eutrophic aquatic environment. The average culturable bacteria grown on the BEP medium were 8.04×10^4, 2.28×10^5, and 6.58×10^8 CFUs per mL or per g dry weight for surface water, bottom water and sediments, respectively, accounting for about 1% of total cells. The percentage was similar to results of the previous studies (Guardabassi et al., 2002).

Table 4.2 Media, dilution factors, organisms and incubation conditions for microorganisms isolated from sediments and freshwater in Yangtze River

Media	Dilution factor[a]	Organisms cultured	Culture conditions	
			Temperature (°C)	Incubation (days)
PTYG	10^4–10^6	Heterotrophic bacteria	28	1–2
R2A	10^0–10^2	Heterotrophic bacteria	28	1–5
Nutrient broth	10^3–10^5	Ammonifying bacteria	28	1–2
Ashby media	10^3–10^5	Nitrogen-fixing bacteria	29	6–7
Alexander-Clark media	10–10^6	Nitrite-oxidizing bacteria	29	25–28
Skinner-Walker media	10–10^6	Ammonia-oxidizing bacteria	29	25–28
Giltey media	10^2–10^8	Denitrifying bacteria and nitrate reducing bacteria	30	2–3
NBRIP organic culture media	10^3–10^5	Organophosphate-solubilizing bacteria	28	2–3
NBRIP media	10^3–10^5	Phosphate-solubilizing bacteria	28	3–5

[a] Dilution factor number is the 1:10 serial dilution from each sample which was plated in triplicate

The nutrient level of the medium exerted a great influence on the count results of culturable bacteria. It was found that the bacterial colony barely grew on the R2A medium that was commonly suggested as the best medium for the cultivation of heterotrophic bacteria in freshwater and soil (Massa et al., 1998). Fig. 4.1 shows that when the nutrient level in the medium was increased, the cultured colonies from freshwater and sediments increased. It was previously reported that the growth of heterotrophic bacteria that lived in a low-nutrient environment (e.g. soil, freshwater and sediment) on a low-nutrient medium was much more vigorous than that grown on a nutritionally rich medium (Balkwill and Ghiorse, 1985; Massa et al., 1998). The possible reason for this phenomenon was that the microorganisms in the eutrophic environment had adapted to the rich nutrient environment, inducing most of those which could only grow on a rich-nutrient medium. Fig. 4.1 also shows that the numbers of bacteria including total bacteria and culturable bacteria in XJ and XX were higher than those in DNs. This was positively related to the physicochemical result which indicated that the nutrient loadings in XJ and XX were higher than those in DNs. We could draw the conclusion that the higher the trophic condition of the water, the more the bacteria that existed.

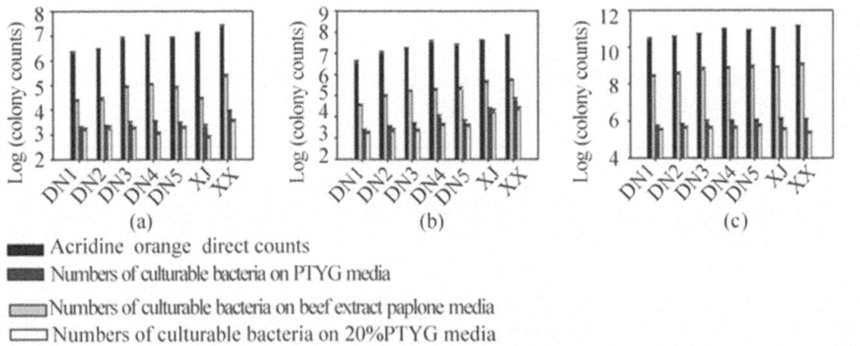

Fig. 4.1 The numbers of total bacteria and culturable bacteria on mediums with different nutrient levels. (a) Surface water; (b) Bottom water; (c) Sediment

4.2.3 Microbial Community Activity

The even activity of the microbial community was investigated by enzyme-linked immunosorbent assay (ELISA) reaction in Biolog ECO microplates, and described using Average Well Color Development (AWCD). Fig.4.2 shows that the activities of microbes increased with the prolonging of incubation time, and the microbial ability utilizing a single carbon source at different sampling sites at the same level was XX>XJ>DN4–DN5>DN3>DN2–DN1. However, among different levels (surface water or bottom water or sediment) the microbial activities were obviously different, and the activity strength was: surface water<bottom water< sediment.

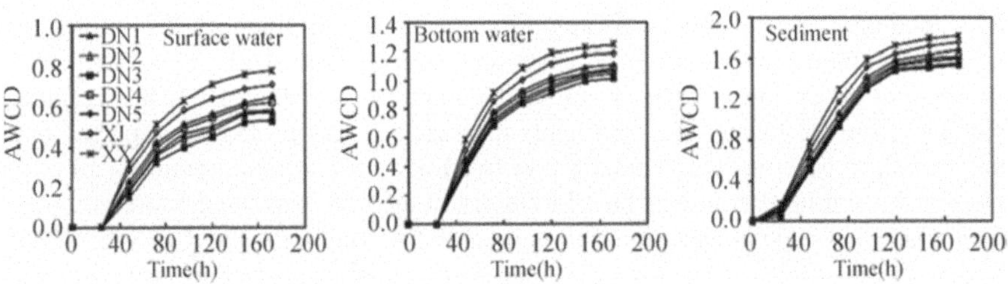

Fig. 4.2 Average well color development (AWCD) of microbial community in surface water, bottom water and sediment

4.2.4 Abundance of Functional Bacteria in Aquatic Environments

Eight kinds of N and P associated functional bacteria, namely AB, NFB, AOB,

NOB, NRB, DNB, OPB and IPB, in surface water, bottom water and sediments were counted separately and the results were plotted in Figs. 4.3 and 4.4.

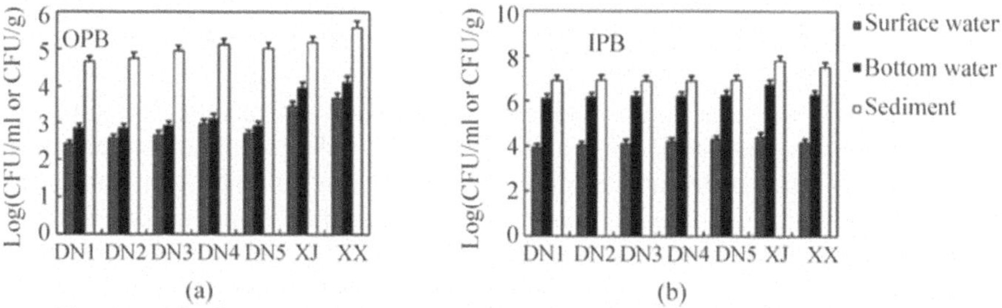

Fig. 4.3 Distribution characteristics of N-associated functional bacteria in water body

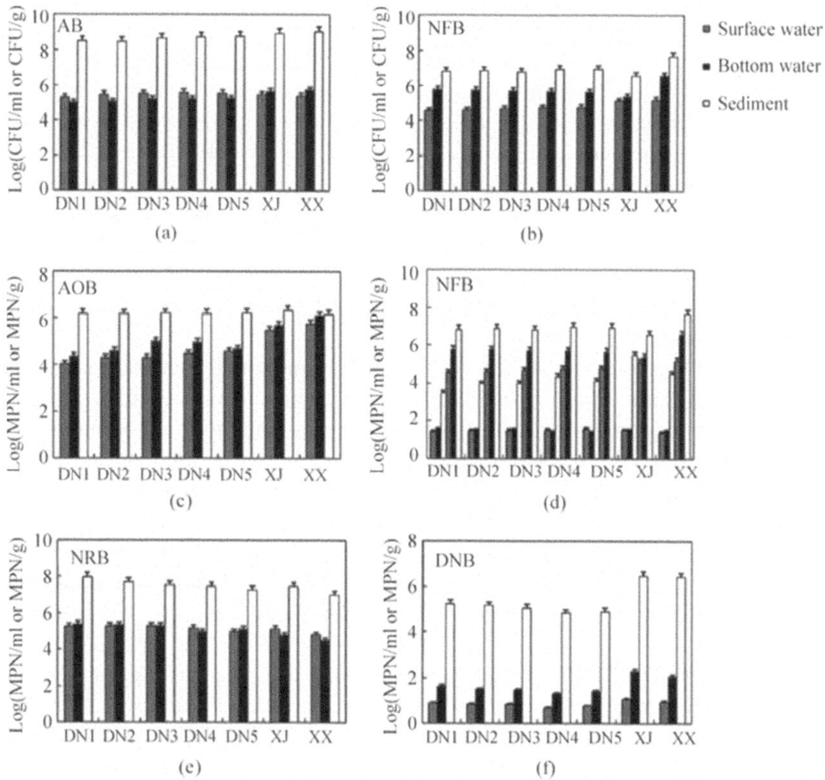

Fig. 4.4 Distribution characteristics of P-associated functional bacteria in water body

As a decomposer, AB can degrade the organic contamination to ammonia, and its abundance can reflect the degree of organic pollution in the water body (Edwards et al., 2001). In the TGR the richness of AB was very high, and its average number of cells were 3.50×10^5, 1.72×10^5, 6.86×10^8 CFUs per mL or per g dry weight in surface water, bottom water and sediments, respectively. The abundance of AB indicated that the organic pollution was rather severe in this aquatic environment (Edwards et al., 2001).

NFB is a kind of bacteria that can transform the nitrogen to ammonia. Fig. 4.3 shows that NFB was abundant in surface water, bottom water and sediments, which were 2.44×10^5, 8.83×10^5 and 1.48×10^7 CFUs per mL or per g dry weight, respectively. This was inconsistent with the research of Van et al. (1998), who stated that NFB was relatively abundant in aquatic environments that lack nitrogen. The reason for the high NFB concentrates was that, in the process of our counting experiment, 0.1 mL of freshwater dilution containing a little nitrogen was added into the Ashby nitrogen-free medium, which could support some bacteria without nitrogen-fixing ability living on the Ashby medium. So, based on this experiment, it was difficult to deduce that this aquatic ecosystem lacked nitrogen.

Nitrification in freshwater, a key process in the nitrogen cycle, requires the coupling of two transformation stages: first, an ammonia-oxidizing stage that transforms NH_4^+ to NO_2^- by AOB, and second, the nitrite-oxidizing stage that transforms NO_2^- to NO_3^- by NOB (Stief et al., 2003). Results showed that the abundance of AOB in bottom water and sediments was high, 5.10×10^4 MPN per mL and 2.40×10^5 MPN per g dry weight, respectively, and the abundance of AOB in surface water was low, which was only about 2.22×10^2 MPN per mL. The NOB richness was much higher than the AOB in this aquatic environment, which was 4.49×10^2, 7.16×10^5 and 3.48×10^7 MPN per mL or per g dry weight in surface water, bottom water and sediments, respectively. The AOB and NOB level revealed in this study could be explained by the fact that nitrification was known to take place mainly on suspended particles and bed sediments (Bonnet et al., 1997; Gresikowski et al., 1996).

NRB and DNB are two kinds of important functional bacteria in the self-purification of water bodies, especially DNB, which can transform inorganic nitrogen-containing compounds to gaseous nitrogen-containing compounds (N_2O or N_2) (Siemens et al., 2003). In these locations, the NRB abundance was very high, generally at a magnitude of 10^5 MPN per mL in water and 10^8 MPN per g dry weight in sediments. Meanwhile, the abundance of DNB was low, which was approximately 7 MPN per mL in surface water and 66 MPN per mL in bottom water. But the abundance of DNB was very high in sediments, approximately 8.75×10^5 MPN per g dry weight, which was mainly due to the anaerobic denitrification.

The enumeration of OPB (Fig. 4.4) showed that these bacterial groups were at the same magnitude of 10^3 CFUs per mL in surface water, bottom water, but they were very abundant in sediments, which were at a magnitude of 10^5 CFUs per g dry weight. The abundance of IPB was 1.51×10^4, 2.24×10^6 CFUs per mL, and 1.97×10^7 CFUs per g dry weight in surface water, bottom water and sediments, respectively. For OPB and IPB, the numbers of cells in the TGR were more than a small eutrophic lake, in which it had been reported that the culturable OPB and IPB were 13 and 89 CFUs per mL in the water bodies, respectively, and about 10^2 and 10^4 CFUs per dry weight in sediments, respectively (Wu and Zhou, 2005). It can be concluded from this study that the abundance of functional bacteria in the eutrophic river was rather high due to the ample nutrient elements: N and P.

4.3 Transformation of Nitrogen

Nitrogen is one of the most important macro nutrients for aquatic organisms. However, discharge of nitrogen to receiving waters can lead to significant impacts on water quality which, with high levels of phosphorus, can cause excessive growth of phytoplankton and eutrophication of water bodies. Nitrification (the oxidation of ammonia to nitrate via nitrite) and denitrification (the reduction of nitrate to molecular nitrogen via nitrite, nitric oxide and nitrous oxide) are essential steps in the water body nitrogen cycle. Nitrification, microbially-mediated oxidation of ammonia to nitrate, is a key process in N cycling in freshwater. It is a two-step process, firstly an ammonia-oxidizing stage that transforms NH_4^+ to NO_2^- by ammonia-oxidizing bacteria (AOB), and secondly the nitrite-oxidizing stage that transforms NO_2^- to NO_3^- by nitrite-oxidizing bacteria (NOB), in which the ammonia-oxidizing stage is the critical stage controlling the nitrification (Bodelier et al., 1996; Costa et al., 2006). In developing a model framework, it is often assumed that nitrification occurs in the water column and that the process follows first-order kinetics with rates calculated as a function of water column ammonia contents (Ambrose et al., 1993; McCutcheon, 1987; Scott and Abumoghli, 1995). However, nitrification in freshwater is known to take place mainly on suspended particles (SPs) and bed sediments (Bonnet et al., 1997; Gresikowski et al., 1996; Pauer and Auer, 2000). Consequently, the investigation of the relationship between SPs and the nitrification rate in the aquatic environment is of importance for developing the ecosystem model of nitrogen biogeochemical cycling (Kittiwanich et al., 2007).

Ammonium ion (NH_4^+), an electrically positive ion species, is very easily adsorbed onto solids. Thus, the adsorption of ammonia on SPs and sediments is also a critical factor to influence the ammonia ion fluxing into the water body. In this context, SPs and sediments play important roles in the sequestration and biogeochemical process of nitrogen (Liu et al., 2007). Hence, analysis of the effects of SPs and sediments on the ammonia adsorption and nitrification in the water body is of great importance for assessing transport, transformation, and the fate of nitrogen on aquatic systems.

Previous studies have separately investigated the adsorption of ammonium ion onto the sediments (Simon and Kennedy, 1987) and nitrification taking place on sediments (Magalhaes et al., 2005). However, no research has focused on simultaneous ammonia adsorption and nitrification. The construction of the TGR has changed the characteristics of SPs in the water body, and consequently the adsorption of ammonia and nitrification taking place on the SPs (Yang et al., 2002). The effects of SPs on the adsorption and nitrification of the nitrifier in a water column were poorly understood till now. The project of this study is to systematically evaluate the influences of suspended particles on the two ammonia biogeochemical processes occurring in the TGR: adsorption and nitrification. As the nitrifiers are composed of two kinds of bacteria species (AOB and NOB), the commonly controlling-rate step, namely the ammonia-oxidizing stage induced by AOB, is used to reflect the nitrification rate. At a constant temperature and with the use of sediments prepared with different treatments, the study aims to explore the influence of various factors including sediment concentration, particle size fraction and organic matter content of sediments on the ammonia oxidation.

4.3.1 AOB Strain and Preparation of Inocula

The culture medium used for the AOB contained 3.8 mmol/L $(NH_4)_2SO_4$, 10 mmol/L NaCl, 1 mmol/L KCl, 0.2 mmol/L $MgSO_4$, 0.1 mmol/L KH_2PO_4, 1 mmol/L $CaCl_2$, and 1 mL trace elements solution (Koops and Moller, 1992) per liter deionized water (Millipore: MilliQ). The sediment samples from which AOB was isolated were collected from the TGR. The preparation stages of the sediment suspension followed those described by Satoh et al. (2003). The AOB in a pure form was isolated and subjected to the following process: The sediment suspension (10 mL) and freshwater were inoculated into a 300-mL Erlenmeyer flask containing 100 mL of AOB culture medium. Shaking (120 r/min) of the medium was continued at 28 °C for 10 d. Five mL of liquor from each culture that produced more than 100 μg/mL of nitrite was inoculated into each flask containing 100 mL of AOB medium and subcultured for 7–10 d. Subculturing was repeated five times. The final subculture was diluted and plated on gellan gum plates and incubated at 28 °C for 10 d. On the plates at dilutions of 10^{-4} and 10^{-5} CFU/mL, colony formation was noted after 14 d of incubation at 28 °C. The colonies from each plate were inoculated into a test tube containing 5 mL of the medium and shaken at 120 strokes/min. After 10–15 d incubation, the isolate that showed the highest growth was noted as SW16. The purification of the strain SW16 was conducted using three types of heterotrophic mediums. The identification process of the strain SW16, including DNA extraction, PCR amplification, the sequencing of amplified 16S rRNA genes and data analysis for construction of a phylogenetic tree, was performed in the previous report (Takahashi et al., 2001).

Cells of the strain SW16 were harvested in the late growth phase from cultures grown in standard media for AOB (Koops and Moller, 1992). Cell suspensions of strain SW16 were obtained by centrifugation, followed by washing and resuspension in sterile saline solution.

4.3.2 Sample Preparation

Sediments were collected from the Three Gorges Reservoir with a Van Veen stainless steel grab sampler (Eijkelamp, Netherlands) in 2007. These freeze-dried sediment samples were ground, homogenized, and stored at −20 °C prior to analysis. To test the effect of particle size distribution on ammonia oxidation, the sediments were size fractionated using the wet-sieve method in the following fractions: >200 μm, 125–200 μm, 63–125 μm, 25–63 μm, and <25 μm. The dependency of ammonia oxidation on the contents of organic matter deposited on sediments was evaluated. Sediments with different organic matter contents were prepared by mixing the raw sediments and the organic-removed sediments according to the given mixing ratios of 0:3, 1:2, 2:1 and 3:0, yielding sediments with organic matter contents of 0.33%, 0.45%, 0.56% and 0.68%, respectively. The organic matter was removed from original sediments by using the H_2O_2 removal method described previously by Wang et al. (2008). After slow addition of 10 mL 30% H_2O_2 solution, a 20 g sediment sample was incubated at 40 °C with intermittent agitation followed by evaporating to dryness. To remove organic matter thoroughly, the above steps were repeated three times.

4.3.3 Analysis and Enumeration

Total organic carbon (TOC) was determined in sediment samples that were acidified with 1.6% HCl and stored at –20 °C (Liqui TOC analyzer, Elementar, Hanau, Germany) (Feng et al., 2007). Ammonia and nitrite were determined with reagent kits for photometric analysis (Merck-Spectroquant®, Merck, Darmstadt, Germany). AOB densities were estimated in water systems with the most-probable-number (MPN) techniques (5 test tube test) using microtiter plates (Lipponen et al., 2002). Although the MPN methodology has been criticized for underestimating the bacteria population (Belser, 1979), it enables a comparison of the potential for nitrite oxidization in different water samples.

To investigate the influence of suspended particles on the nitrification, rates of ammonia oxidizing were calculated by observing changes in concentrations of NO_2^- over time. The original NH_4Cl concentration in treated surface water was ~3 mg N/L. After adding a given mass of sediment samples and 300 mL water sample in 500 mL of conical flasks, the microcosms were inoculated with 2.5 mL of inoculum and incubated at 180 rpm in a constant temperature shaking incubator (T = 27 °C) in the dark for 8 d. In preliminary experiments, inspection of the ammonia and nitrite data indicated that nitrification resulting from physicochemical reaction was negligible. Six milliliters of aliquots were removed from the microcosms once daily, in which 1 mL was used to analyze the density of AOB cells, and 5 mL underwent nitrite-nitrogen content determination after centrifuging. The centrifuged sediment pellet was flushed with equal deionized water and added to the microcosms replacing the removed water to maintain a constant volume of water and sediment.

4.3.4 Ammonia Nitritation

Because the nitrification involves two kinds of nitrifiers, AOB and NOB, it is difficult to evaluate the individual contribution of AOB and NOB during the nitrification process. The step controlling the nitrification rate, namely the ammonia-oxidizing stage induced by AOB, was chosen to reflect the nitrification rate. As thus, an AOB strain SW16 was isolated from sediments collected from the TGR. An almost-complete 16S rRNA gene sequence of strain SW16, containing less than 1% undetermined positions, was obtained. NCBI BLAST, the Basic Local Alignment Search Tool (BLAST) is a suite of programs designed to search all available sequence databases for similarities between a protein or DNA query and known sequences. In this study, BLAST was used to match the sequence relationships, providing scores that can distinguish real matches from background hits with a high degree of statistical accuracy. The result of the BLAST program indicated that the isolate possessed a 16S rRNA gene sequence with 98% similarity to that of the species Nitrosomonas nitrosa in GenBank, hence AOB strain SW16 was classified as Nitrosomonas nitrosa. Cell suspensions of strain SW16 used in this study was diluted to a concentration of approximately 2.3×10^8 cells per milliliter of solution (SE 0.063×10^8).

4.3.5 Influence of Suspended Particle Concentration on Nitritation

The nitritation experiments under different suspended particle concentrations were carried out using AOB strain SW16. The preliminary experiments revealed that the ammonia was completely depleted within 8 days under typical environmental conditions in the TGR (suspended particle concentration was 2 g/L and ammonia concentration in supernatant was 0.2 mg N/L), so that in this test the nitritation experiment was performed for 8 days. Fig. 4.5 depicts the nitrite productions coupled with the ammonia oxidation. In the first 2 days there was no nitrite production, whereas in the following 6 days the nitrite production dramatically increased. The biological transformation was completed in 8 days with the presence of sediments, but only 34% of ammonia was transferred to nitrite with the absence of sediments. The rate of nitrite production was enhanced with the increase in the suspended particle concentrations (Fig. 4.5). For example, the transformation rate of ammonia to nitrite in 6 days approached 40%, 52%, 76% and 91% with suspended particles of 1, 2, 5 and 10 g/L, respectively. This suggests that the presence of suspended particles facilitates the nitrogen biological transformation, and the increasing SP concentration accelerates the nitritation process.

The AOB cell concentrations in sediment-water solutions were audited during the 8-day nitritation process, as shown in Fig. 4.6. The initial AOB densities, ca. 1.2×10^6 cells/mL, were similar in solutions with different suspended particle concentrations with the addition of strain SW16 inoculum (1 mL). After the lag phase of 2 days, the AOB growth entered into the logarithmic phase, and then reached the stationary phase with the consumption of ammonia. Fig. 4.6 shows that the growth rates of AOB were higher in the higher suspended particle concentrations. Specifically, after 6-day ammonia oxidation, the AOB densities reached 3.20×10^6, 9.13×10^6, 10.5×10^6, 12.6×10^6 and 13.6×10^6 cells/mL at SP concentrations of 0, 1, 2, 5 and 10 g/L, respectively. This can explain the observation made above that the rate of nitrite production improved with the increase in suspended particle concentrations. Moreover, it can be seen from Fig. 4.5 and Fig. 4.6 that there is a significant correlation (*t*-test, $P<0.05$) between the ammonia oxidizing rate and the AOB growth rate.

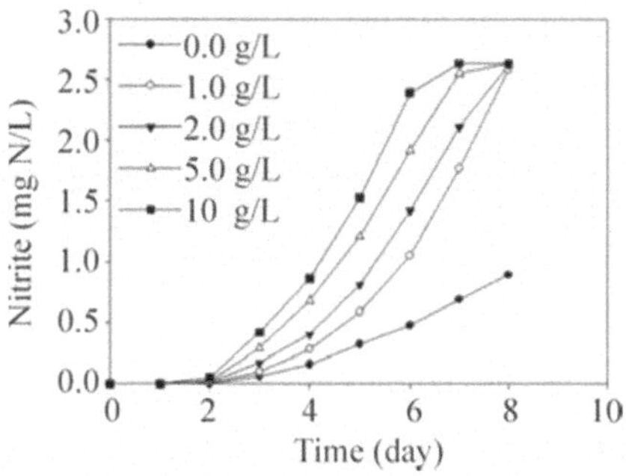

Fig. 4.5 Effect of suspended particle content on the nitritation (T=28 °C, initial NH_4^+-N concentrations =2.63 mg/L)

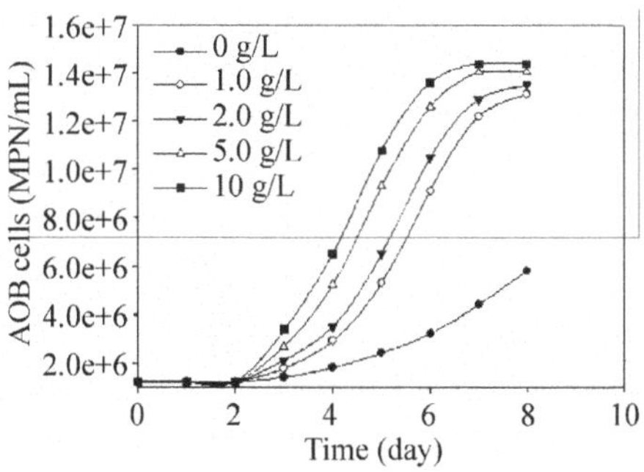

Fig. 4.6 The counts of ammonia-oxidizing bacteria (AOB) with different concentrations of suspended particles (T=28 °C, initial NH_4^+-N concentrations =2.63 mg/L)

Many groups of bacteria have been shown to exist predominantly as attached colloids on surfaces in contact with liquids (Davies, 2000; Palmer et al., 2007). The advantages gained by the living bacteria attached to a surface are thought to improve the growth rate and activity of the microorganisms by the higher concentration of nutrients close to a surface and promote genetic exchange (Donlan, 2001). Moreover, there are many claims that surface attachment appears to protect nitrifying bacteria from a range of inhibitors (Foppen et al., 2008; Marina et al., 2000; Ng and Stenstrom, 1987; Park et al., 2003). Marina et al. (2000) and Ng and Stenstrom (1987) reported that the addition of powdered activated carbon to an activated sludge wastewater treatment process enhanced nitrification by adsorbing inhibitory compounds. Park et al. (2003) reported that the enhanced nitrification efficiency of activated sludge with the addition of powder activated carbon or zeolite was accomplished by the attached growth of nitrifier on the surface of carriers. The mechanism of the improvement in the nitrification rate with the addition of nitrifier carriers is still not well understood, but the surface growth of nitrifier may be a factor affecting the nitrification efficiency. Previous researchers have found that the nitrifying bacteria in the water column of the lake and river were low, with low or even no observed nitrification in the water column (Cirello et al., 1979; Hall, 1982; Pauer and Auer, 2000). This is attributed to the low nitrifier biomass concentrations and the low growth rates of AOB in the water column, probably as a result of a low level of SPs which were considered as a main habitat for the biological nitrification in the water body (Bonnet et al., 1997; Gresikowski et al., 1996; Pauer and Auer, 2000).

4.3.6 Influence of Particle Size and Organic Matter Content on Ammonia Oxidation

To investigate the influences of suspended particle size and organic matter content on the microbial-nitrogen cycle in typical Yangtze River conditions, the experiments of nitration in water-SP solution with different particle sizes and organic matter content SPs were performed. Fig. 4.7 plots the nitrite production with different particle sizes, showing that the influence of particle size is not obvious in experimental conditions. Specifically, for suspended particle size fractions of >200, 125–200, 63–125, 25–63, and <25 μm, the nitrite production was 1.73, 1.76, 1.77, 1.75 and 1.82 mg N/L, respectively in 6 days. Previous studies (Ling et al., 2002; Muirhead et al., 2006; Oliver et al., 2007) found that bacteria (*E. coli*) would associate with soil particles and, furthermore, cells would preferentially attach to a particular soil particle size fraction. It was found that E. coli preferentially attached to the soil particle size fraction of 30−16 μm when the observation was made with the soil particle size fractions of >31, 30–16, 15–4, 3–2 and <2 μm, respectively. However, in this study, the discrimination of nitritation resulting from AOB attachment on the different size fractions was negligible to observe. The mechanism involved in the discrepancy between this study and previous reports is not clear, providing a further issue of concern.

Fig. 4.8 indicates the nitritation process of AOB under different organic matter contents containing in suspended particles. The ammonia oxidizing rates of AOB with different organic matter content sediments were not significant. Nevertheless, other researchers found that organic matter is frequently cited as a factor that limits nitrification in activated sludge (Abeliovich, 1992; Foppen et al., 2008; White and Gosz, 1987). This probably happens due to the fact that under the experimental conditions the organic matter content contained in SPs is low, and their discrimination is not sufficient to cause the change in the nitritation process.

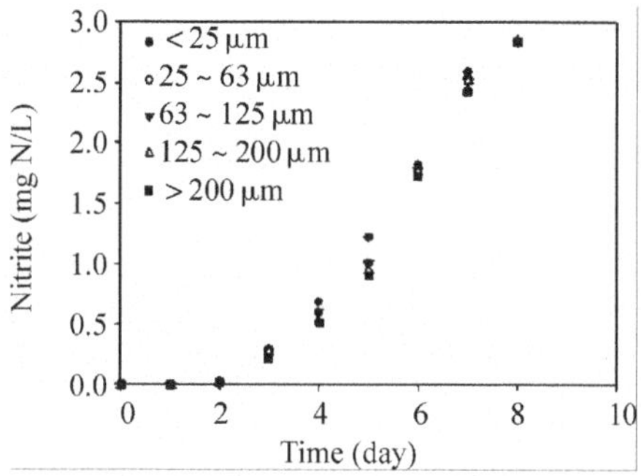

Fig. 4.7 Effect of suspended particle size on the nitritation (T=28 °C, sediment concentration=2 g/L, initial NH_4^+-N concentrations =2.84 mg/L)

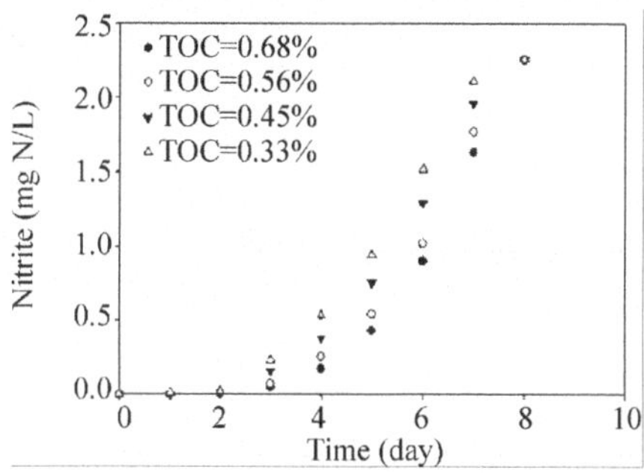

Fig. 4.8 Effect of organic matter contained in suspended particles on the nitritation (T=28 °C, sediment concentration=2 g/L, initial NH_4^+-N concentrations =2.26 mg/L)

4.4 Transformation of Phosphorus

As reported, phosphorus in sediments exists in several forms of organic phosphorus (OP), water-soluble inorganic phosphate and water-insoluble inorganic phosphorus (such as calcium bound P and Fe/Al bound P) (Golterman, 1996; Ruban et al., 2001). Wherein, water-soluble inorganic phosphate is the most important one because it can be taken up by phytoplankton directly and therefore contributes to water body productivity more effectively. Water-insoluble inorganic phosphorus and OP usually cannot be utilized by phytoplankton directly, but can be transformed to soluble forms under physical, chemical and biological pressures (Evans and Johnes, 2004; Wu and Zhou, 2005).

Phosphorus release from sediments to water is a complicated process which is influenced by many factors. So far, the physical effect (such as temperature, disturbance) and chemical effect (such as the change in acid-alkali and oxidation-reduction) have been studied at length for many years (Kim et al., 2003; Gardolinski et al., 2004). However, few attempts were made on microorganisms with the ability to dissolve water-insoluble inorganic phosphorus or mineralize OP, which has been regarded as an important influencing factor for P release (Kim et al., 2005).

Recently, the effect of phosphate-solubilizing bacteria (PSB) on P release from sediments to water has attracted more and more attention (Wu and Zhou, 2005). Studies indicated that the solubilization of water-insoluble inorganic phosphorus by PSB was associated with the production of organic acid which may chelate the cations bound to phosphate, thereby converting it into soluble forms to release (Chen et al., 2006). The mineralization of OP by PSB depends on enzymolysis of alkaline phosphatase (AKP) secreted by microorganisms (Mhamdi et al., 2007). Many factors, such as nutritional, physiological and growth conditions, influenced the ability PSB for phosphate-solubilizing. Inversely, PSB may change the environmental conditions (such as pH, DO) by metabolism, and thus influence the P transfer from sediment to water. That is, P release from sediment to water is not a simple process, especially in the condition of existing bacteria. Thus it is necessary to further study the effect of PSB on P release.

4.4.1 Samples Characteristics

The characteristics of sediment and surface water samples from the Daning River were summarized in Table 4.3. Organic matter, expressed as TOC and LOI, showed the lowest value of 0.58% and 3.85% in Daning C sediment, the highest of 2.19% and 7.23% in Daning B sediment. Moreover, the highest contents of TP, Org-P and Inorg-P occurred in Daning B sediment. It indicated that Daning B sediment has the highest value of TOC, LOI, TP, Org-P and Inorg-P, which may relate to the fact that a relatively large amount of human waste was discharged around this sample site The contents of TP in five sediment samples ranged from 0.416 to 1.184 mg/g. Wherein the contents of Inorg-P and Org-P were 0.184–0.691 mg/g (42.2%–58.4% of TP) and 0.170–0.478 mg/g (38.5%–45.9% of TP), respectively. This result suggested that the content of Inorg-P was higher than that of Org-P in Daning sediments except for Daning D. Inorg-P of the sediments included mainly Fe/Al-P (0.031–0.249 mg/g) and Ca-P (0.07–0.481 mg/g). The ratio of Fe/Al-P and Ca-P was 0.253–0.786, indicating Ca-P was the dominant existing form in Inorg-P for the Daning sediments.

As shown above, Ca-P and Org-P were 2 principal phosphorus-existing forms in Daning sediments (Table 4.4). So PSB for solubilization of Ca-P (IPB) and mineralization of Org-P (OPB) may be existed in samples. In our study, Daning C was used as an example for the enumeration of PSB using the conventional dilution plate-count method. The results revealed that the counts of IPB grown on calcium phosphate agar were about $(848.20\pm137.45)\times10^4$ CFU/g in sediment and $(1.27\pm0.18)\times10^4$ CFU/mL in water, which were obviously higher than that of OPB grown on phosphatidylcholine agar ($(8.56\pm0.64)\times10^4$ CFU/g in the sediment, $(0.047\pm0.00057)\times10^4$ CFU/mL in water). That is to say, more IPB than OPB was observed in both sediment and water. As reported, microorganisms would be easier to grow on the sediment where more abundant nutrients (C, N and P) existed than in water (Wu and Zhou, 2005). So the amount of PSB in the sediment was apparently higher than that in water, which was also obtained in our study.

Table 4.3 The main characteristics of sediment and surface water

Sample	Sediment characteristics							Surface water characteristics			
	TOC (%)	LOI (%)	TP (mg/g)	Org-P (mg/g)	Inorg-P (mg/g)	Fe-P (mg/g)	Ca-P (mg/g)	TP (mg/L)	TDP (mg/L)	PO_4^{3-} (mg/L)	TN (mg/L)
DaningA	0.960	4.313	0.827	0.320	0.349	0.074	0.171	0.057	0.037	0.014	1.048
DaningB	2.190	7.230	1.184	0.478	0.691	0.121	0.481	0.078	0.071	0.005	1.772
DaningC	0.580	3.847	0.903	0.386	0.513	0.249	0.316	0.077	0.073	0.0003	2.025
DaningD	1.540	6.190	0.416	0.191	0.184	0.031	0.070	0.089	0.032	0.006	3.807
DaningE	0.720	4.070	0.442	0.170	0.211	0.031	0.098	0.062	0.045	0.020	2.294

4.4.2 The Phosphorus Release Ability of PSB

In this chapter, by isolation and purification, 8 predominant IPB and 3 OPB were obtained to evaluate phosphorus release abilities. Fig. 4.9a showed the variation of the accumulated PO_4^{3-}-P concentration and pH in the medium after inoculating with IPB6. Obviously, the accumulated PO_4^{3-}-P concentration in the medium increased before 60 h, accompanied by a sharp decrease in pH of the liquid (from 7.01 to below 5.0). This result was also obtained by other authors (Kim et al., 2005). Then few variations occurred in PO_4^{3-}-P concentration and pH of the liquid. Comparably, OPB1 mineralized phosphatidylcholine weakly, accumulated a small amount of soluble PO_4^{3-}-P (Fig. 4.9b). After 36 h of inoculation, PO_4^{3-}-P concentration just began to increase, which may suggest that OPB1 grew slowly. Additionally, no apparent decrease appeared in the pH of the liquid.

It is generally accepted that the mechanism of phosphorus solubilization by IPB was associated with the production of acidic metabolites by bacteria, which debase pH in the liquid and chelate the cations bound to phosphorus through their hydroxyl and carboxyl groups and then release PO_4^{3-}-P into the liquid phase (Kim et al., 2005; Chen et al., 2006). Additionally, the release of protons accompanying respiration or ammonium assimilation was another probable reason for phosphorus solubilization (Illmer and Schinner, 1995; Reyes et al., 1999). As for OPB1, pH of the medium was still as high as 7.36 after 7 days incubation, suggesting that other mechanisms of phosphorus solubilization should exist. It was well known that most of OP cannot be assimilated directly by microorganisms but can be transformed to soluble inorganic P by microbial enzymes such as AKP. In our study, when phosphatidylcholine is the sole P source, AKP may be secreted by OPB to mineralize OP (Sahu and Jana, 2000). The more AKP secreted by OPB, the higher ability for OP mineralization. Therefore, both the accumulation of PO_4^{3-}-P in the liquid and the activity of AKP can be used to evaluate the OP mineralization ability of OPB.

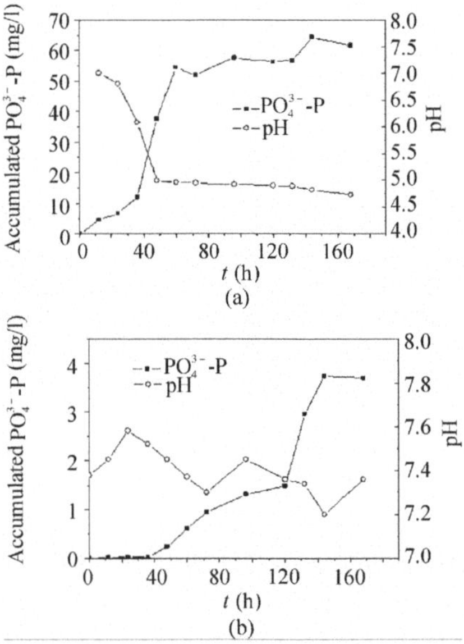

Fig. 4.9 Temporal changes of PO_4^{3-}-P and pH in liquid medium. (a) The ability of IPB6 for calcium phosphorus solubilizing; (b) The ability of OPB1 for phosphatidylcholine mineralization at shake culture (30 °C, 150 r/min)

Table 4.4 summarizes the details of pH and the amounts of soluble-P in the medium after 72 h and 144 h of incubation by different PSB. The soluble-P concentration from TCP in the medium ranged from 14.71 mg/L to 64.39 mg/L, which was accompanied by a drop in pH (4.82–6.69) from an initial pH of 7.01 after 144h. The maximum P solubilization was recorded by IPB6 with a maximum drop in the pH to 4.82. Comparably, the release of soluble-P from phosphatidylcholine mineralization by OPB was only from 1.97 mg/L to 3.73 mg/L with no obvious variation in pH. The activities of AKP were between 38.63 mg/L and 62.94 mg/L, demonstrating AKP was secreted by OPB during the period of incubation. Wherein, the higher ability of AKP obtained by OPB1 accelerated the mineralization of OP, and thus more soluble-P ions were released. To sum up, IPB6 and OPB1 showed the highest activity for IP solubilization and OP mineralization in our study, respectively. So they were selected to study the effect of PSB on P release from Daning C sediment.

Table 4.4 pH, the concentration of PO_4^{3-}-P in the medium and the activity of AKP secreted by OPB after 72 h and 144 h of incubation for different PSB (30 °C, 150 r/min)

Isolates	Time	IPB1	IPB2	IPB3	IPB4	IPB5	IPB6	IPB7	IPB8	OPB1	OPB2	OPB3
pH	72 h	6.38	5.53	7.06	5.22	5.46	4.95	5.39	4.98	7.32	7.24	7.31
PO_4^{3-}-P (mg/L)	72 h	17.48	24.75	6.38	36.16	29.15	52.02	27.32	50.67	0.95	0.47	0.82
AKP (mg/L)	72 h	/	/	/	/	/	/	/	/	56.19	32.31	48.84
pH	144 h	5.79	5.41	6.69	5.16	5.60	4.82	5.37	5.01	7.20	7.31	7.26
PO_4^{3-}-P (mg/L)	144 h	40.00	34.49	14.71	45.33	33.05	64.39	31.78	57.98	3.73	1.97	3.36
AKP (mg/L)	144 h	/	/	/	/	/	/	/	/	62.94	38.63	55.97

4.4.3 Release of Phosphorus from Sediment Using PSB at Different Temperatures

The release of PO_4^{3-}-P from Daning C sediment in a slurry reactor which was inoculated with IPB6 or OPB1 was shown in Fig. 4.10. The experiment with no effect of bacteria by adding chloroform was used as control. When the activity of bacteria was inhibited, the concentration of PO_4^{3-}-P in liquid increased with time, indicating that P transfers from the sediment to water at the shaking condition. The maximum PO_4^{3-}-P concentration in liquid was 0.081, 0.145 and 0.275 mg/L at 10 °C, 20 °C and 30 °C, respectively. Then PO_4^{3-}-P concentration decreased slightly and kept steady at different temperatures.

After PSB was introduced into the reactor, no obvious difference in phosphate concentration was observed at the initiation of the experiment compared with the control. But after 1 day, higher phosphate release concentration occurred in the reactor because of the effect of bacteria. As for IPB6, the phosphate concentration in water enhanced continuously, reaching 0.123, 0.199 and 0.419 mg/L after 9 days at 10 °C, 20 °C and 30 °C, respectively. These values were about 0.61–1.02 times higher than those in the control. Additionally, no apparent change in pH was observed because of the strong buffering capacity of the sediment. Comparably, the phosphate release concentration also rose in the reactor introduced to OPB1, but then a decreasing trend appeared after 6 days. The maximum phosphate concentrations achieved 0.091, 0.172 and 0.357 mg/L at 10 °C, 20 °C and 30 °C, respectively, which were lower than those in the reactor introduced to IPB6. The changing trend in AKP activity was about consistent with that of the phosphate concentration in water, which may be due to the fact that the enzymolysis of AKP was the main reason for OP mineralization.

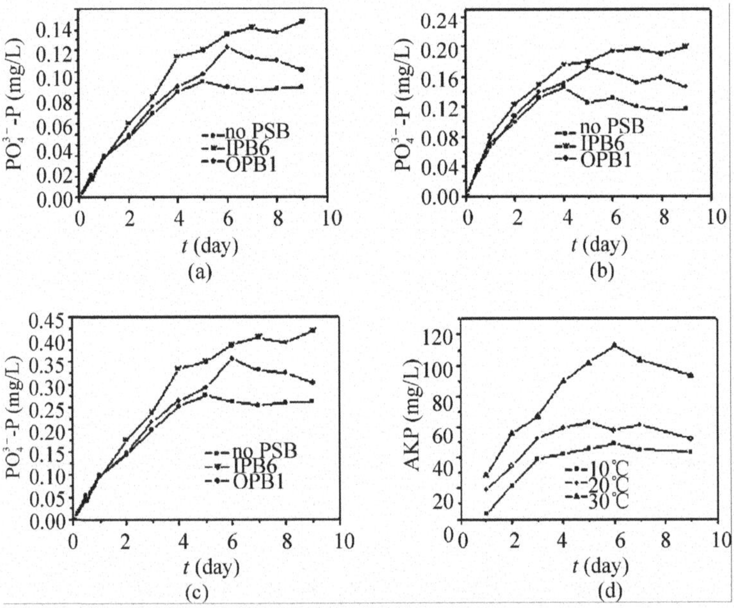

Fig. 4.10 Temporal changes in PO_4^{3-}-P concentration in the aqueous phase of the slurry mixture at (a) 10 °C; (b) 20 °C; (c) 30 °C. (d) The activity changes in AKP secreted by OPB1 over time at different temperatures (10 °C, 20 °C, 30 °C)

The phosphorous release from the sediments to water is not only a physical process but also related to the biological circulation of P. When the activity of bacteria was inhibited, phosphate adsorbed on the surface of the sediment was released quickly under a highly disturbed intensity until a dynamic balance of release and adsorption for P was reached. After PSB was introduced into the reactor, the effect of microbial decomposition for insoluble P in the sediment should be considered for P release. With the growth of IPB and OPB, calcium phosphorus was solubilized and OP was mineralized continuously, which was the reason that higher phosphate concentration was observed in the reactor introduced to IPB6 or OPB1 than in the control. During the course of OP mineralization, the increasing activity of AKP secreted by OPB1 because of the absence of water-soluble inorganic phosphorus accelerated OP mineralization at the initiation of the experiment. That is, more phosphate was released from sediment to water. However, with the increase in phosphate concentration in the water, the activity of AKP was inhibited, which was reported by other researchers (Thomsen et al., 2002; Mhamdi et al., 2007). So lower activity of AKP was observed after a large amount of phosphate was released, which inversely decreased OP mineralization.

Also, Fig. 4.10 indicated that a higher temperature accelerated the P release. The dissolving of phosphate minerals existing in sediments was related to the dissolvability which may be enhanced by higher temperatures. Moreover, the activity of bacteria increased with the rising temperature (from 10 °C to 30 °C), which would accelerate more insoluble P decomposition and then release this from sediment to water.

As shown above, the decomposition effect for insoluble P in sediments by PSB was obvious. That is, some Ca-P and OP in sediment may be decomposed by IPB and OPB, respectively, to form soluble phosphate released to water. P decomposed by PSB was also partly assimilated for the growth of bacteria, and then reserved inside the cell. To clarify the P content and form inside the bacteria, P assimilated by PSB was released ultrasonically. The comparison of TDP and PO_4^{3-}-P concentrations in the aqueous phase after 9 days incubation at different temperatures before and after ultrasonic treatment is shown in Table 4.5. It can be seen that, compared with the control experiment, the concentrations of TDP and PO_4^{3-}-P in the reactor obviously increased after ultrasonic treatment. For example, after ultrasonic treatment, the increment of TDP and PO_4^{3-}-P in the reactor introduced with IPB6 was 0.091 and 0.046 mg/L, respectively, which was 1.84 and 1.42 times higher than that in the control experiment (20 °C). This result demonstrated P decomposed by PSB was partly reserved inside the bacteria as OP and inorganic phosphate form. Thus the P decrement from the sediment was the sum of P release into the water and the P assimilated by bacteria was provided for the growth of bacteria and reserved inside the cells. After these bacteria died, P preserved inside the bacteria would also be released into water.

Table 4.5 The comparison of TDP and $PO4^{3-}$-P concentrations in the aqueous phase of the slurry mixture after 9 days incubation at different temperatures before and after ultrasonic treatment

Temperature (°C)		No PSB		IPB6		OPB1	
		TDP (mg/L)	PO_4^{3-}-P (mg/L)	TDP (mg/L)	PO_4^{3-}-P (mg/L)	TDP (mg/L)	PO_4^{3-}-P (mg/L)
10	A (before ultrasonic)	0.156	0.061	0.242	0.123	0.171	0.072
	B (after ultrasonic)	0.177	0.077	0.317	0.156	0.221	0.103
	B-A	0.021	0.016	0.075	0.033	0.050	0.031
20	A (before ultrasonic)	0.197	0.116	0.310	0.199	0.246	0.145
	B (after ultrasonic)	0.229	0.135	0.401	0.245	0.309	0.182
	B-A	0.032	0.019	0.091	0.046	0.063	0.037
30	A (before ultrasonic)	0.357	0.261	0.545	0.419	0.412	0.303
	B (after ultrasonic)	0.39	0.287	0.696	0.484	0.489	0.357
	B-A	0.033	0.026	0.151	0.065	0.077	0.054

4.4.4 The Effect of DO on Phosphorus Release from Sediment Using PSB

Dissolved oxygen (DO), which denotes the oxide reductive condition in the system, is one of the most important influencing factors for P release from sediment to water. Lower DO made some metals such as Fe, Al in the sediment reduce, and thus induced the release of P bound to these metals. Additionally, the activity and community structure of bacteria in sediment was strongly influenced by DO. That is, the different ability of PSB for phosphorus solubilization would be observed in anoxic conditions. Thus it was necessary to compare the P release from Daning C sediment to water in anoxic conditions with that in aerobic conditions.

To realize P release from Daning C sediment to water in anoxic conditions, nitrogen was introduced into the reactor to strip out oxygen at the initial stage, which made DO in the slurry reactor remain at a lower than 1 mg/L. After 9 days incubation at 30 °C, the concentration of PO_4^{3-}-P in the aqueous phase of the slurry mixture reached 0.607 mg/L for IPB6 introduced, 0.496 mg/L for OPB1 introduced and 0.326 mg/L for the control, respectively, which was obviously higher than that in the aerobic condition, especially for that introduced with PSB (Fig. 4.11). This result indicated that lower DO accelerated P release from sediment to water, which was also obtained by other reports (Kim et al., 2003). Moreover, the existence of PSB may enhance the amount of P release.

According to the result obtained in section 3.3, P decomposed by PSB was not only released to water, but also assimilated by the bacteria to be preserved in the cell. After 9 days incubation in anoxic condition at 30 °C, the P preserved in the cell was released by ultrasonic treatment. The TDP and PO_4^{3-}-P increment, which mainly came from the inside of bacteria, was compared with that in aerobic conditions (Figs. 4.11b and 4.11c). It can be seen that there was a slight increase for the control experiment. However, the TDP and PO_4^{3-}-P increments were 0.118 and 0.046 mg/L for IPB6 introduced, 0.049 and 0.032 mg/L for OPB1 introduced, respectively, which were obviously lower than those in the aerobic condition. This result indicated that lower P assimilated by bacteria in anoxic condition, i.e. lower DO, debased the activity of PSB for P solubilization. So there must be some other reason for higher P release from the sediment to water in anoxic conditions. As reported, Fe(III) was easier to reduce to Fe(II) in anoxic conditions, which thus induced Fe-bound P to release into water. In our study, after 9 days incubation at 30 °C in anoxic conditions, the concentration of Fe ions in water increased obviously by ICP analysis, demonstrating that the release of Fe-P occurred in anoxic conditions. Therefore, it could be inferred that the release of Fe-P in anoxic conditions increased the concentration of phosphate in water. Moreover, Fe(III) in sediment is one of the most favourable electron acceptors for bacteria respiration in anoxic conditions. That is, Fe(III) in sediment was easier to reduce depending on the effect of bacteria, which induced more Fe-P released into water when PSB existed.

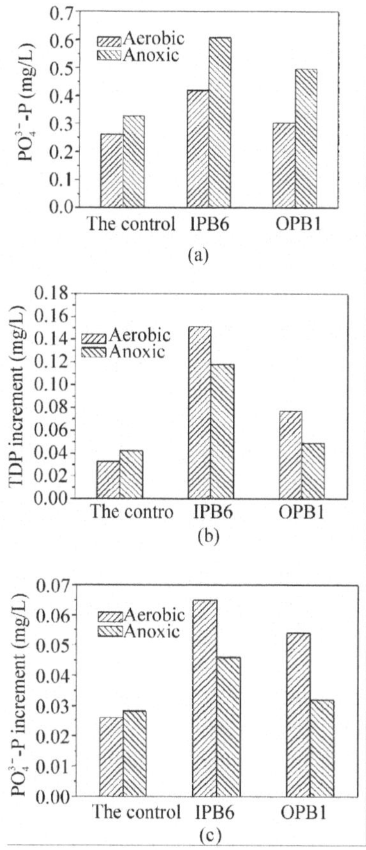

Fig. 4.11 The comparison of TDP and PO_4^{3-}-P concentrations in the aqueous phase of the slurry mixture after 9 days incubation at different temperatures before and after ultrasonic treatment

The aquatic ecosystems of the backwater areas of the Yangtze River in the TGR, where water bloom frequently occurred, were in an eutrophic condition. Eutrophic conditions in the Xiaojiang River and Xiangxi River were more severe than in the Daning River. The total bacteria and culturable bacteria were relatively abundant in the TGR. Our experimental results revealed that the richer the nutrient level of media, the more the culturable bacterial colonies cultured in eutrophic freshwater. The results of bacteria counting and microbial community analysis revealed that the increase of nutrient loadings would give rise to bacteria densities yet decrease bacteria community diversities. Eight kinds of functional bacteria associated with N or P, namely AB, NFB, AOB, NOB, NRB, DNB, OPB and IPB, were found to be abundant in the backwater areas in the TGR.

Laboratory microcosm experiments were used to estimate the influence of suspended particles on the nitritation process. AOB strain SW16, identified as *Nitrosomonas nitrosa*, was used to determine the nitritation process. Results showed that the suspended particles played an important role in the nitritation, and the ammonia oxidizing rate was enhanced with the increase in suspended particle concentration. The significant improvement in nitritation under the high SP concentrations is attributed to the high nitrifier biomass concentrations resulting from the fast growth rates of AOB caused by the high level of SPs, which were considered as the main habitats for the biological nitrification in the water body. Under the experimental conditions, no obvious changes in nitrite production were observed in the nitrification process with the different particle size fractions and the organic matter concentrations loaded on sediments.

The analysis for samples collected from the Daning River, an important tributary of the Three-Gorges Reservoir (China), showed that TP content in Daning sediments ranged from 0.416 to 1.184 mg/g. Wherein, Ca-P and Org-P were 2 principal phosphorus-existing forms, which induced the PSB for solubilization of Ca-P and mineralization of OP and were abundant, especially in sediment.

The comparison of P release ability of IPB and OPB isolated from sediments showed that IPB possessed a higher ability because of the production of acidic metabolites. AKP secreted by OPB for OP mineralization was inhibited by increasing phosphate concentration in water, and thus debased the P release. Laboratory tests on P release from Daning sediment revealed more P was released when PSB existed and, simultaneously, some P was reserved inside the bacteria as OP and in inorganic phosphate form, which would be also released into water after PSB died.

Higher temperatures enhanced the activity of PSB, and thus increased the amount of P release and P reserved inside the bacteria. In anoxic conditions, the amount of P reserved in PSB decreased. However, higher phosphate concentration was observed in water because of the release of Fe-P which was more obvious, depending on the effect of bacteria.

References

Abeliovich A (1992) Transformations of ammonia and the environmental impact of nitrifying bacteria. Biodegradation 3:255-264.

Ambrose RB, Wool T, Martin JL (1993) The Water Quality Analysis Simulation Program, WASP5, Part A: Model Documentation. In: USEPA (ed.), Athens GA: U.S. EPA National Exposure Research Laboratory, Ecosystems Division.

Balkwill DL, Ghiorse WC (1985) Characterization of subsurface bacteria associated with two shallow aquifers in Oklahoma. Appl Environ Microbiol 50:580-588.Belser LW (1979) Population ecology of nitrifying bacteria. Annu Rev Microbiol 33:309-333.

Bodelier PL, Libochant JA, Blom CW, Laanbroek HJ (1996) Dynamics of nitrification and denitrification in root-oxygenated sediments and adaptation of ammonia-oxidizing bacteria to low-oxygen or anoxic habitats. Appl Environ Microbiol 62:4100-4107.

Bonnet C, Volat B, Bardin R, Degrange V, Montuelle B (1997) Use of immunofluorescence technique for studying a nitrobacter population from wastewater treatment plant following discharge in river sediments: first experimental data. Water Res 31:661-664.

Borsodi AKF, Kurdi P (1998) Numerical analysis of planktonic and reed biofilm bacterial communities of lake Ferto. Water Res 32(6):1831-1840.

Chen YP, Rekha PD, Arun AB, Shen FT, Lai WA, Young CC (2006) Phosphate solubilizing bacteria from subtropical soil and their tricalcium phosphate solubilizing abilities. App Soil Ecol 34:33-41.

Cirello J, Rapaport RA, Strom PF, Matulewich VA, Morris ML, Goetz S, Finstein MS (1979) The question of nitrification in the Passaic River, New Jersey: analysis of historical data and experimental investigation. Water Res 13:525-537.

Coolen MJ, Overmann J (2000) Functional exoenzymes as indicators of metabolically active bacteria in 124,000-year-old sapropel layers of the eastern Mediterranean Sea. Appl Environ Microbiol 66:2589-2598.

Costa E, Perez J, Kreft JU (2006) Why is metabolic labour divided in nitrification? Trends Microbiol 14:213-219.

Davies DG (2000) Physiological events in biofilm formation. Society for General Microbiology 37-51.

Donlan RM (2001) Biofilm formation: a clinically relevant microbiological process. Clin Infect Dis 33:1387-1392.

Du LN, et al. (2011) Effect of eutrophication on molluscan community composition in the Lake Dianchi (China, Yunnan). Limnologica 41(3):213-219.

Edwards ML, Lilley AK, Timms-Wilson TH, Thompson IP, Cooper I (2001) Characterisation of the culturable heterotrophic bacterial community in a small eutrophic lake (Priest Pot). FEMS Microbiol Ecol 35:295-304.

Evans DJ, Johnes PJ (2004) Physico-chemical controls on phosphorus cycling in two lowland streams. Part 1-the water column. Sci Total Environ 329:145-163.

Feng JL, Yang ZF, Niu JF (2007) Remobilization of polycyclic aromatic hydrocarbons during the resuspension of Yangtze River sediments using a particle entrainment simulator. Environ Pollut 149:193-200.

Foppen JW, Liem Y, Schijven J (2008) Effect of humic acid on the attachment of Escherichia coli in columns of goethite-coated sand. Water Res 42:211-219.

Gardolinski PCFC, Worsfold PJ, McKelvie ID (2004) Seawater induced release and transformation of organic and inorganic phosphorus from river sediments. Water Res 38:688-692.

Gasol JM, Duarte CM (2000) Comparative analyses in aquatic microbial ecology. FEMS Microbiol Ecol 31:99-106.

Golterman HL (1996) Fractionation of sediment phosphate with chelating compounds: a simplification, and comparison with other methods. Hydrobiologia 335:87-95.

Gresikowski S, Greiser N, Harms H (1996) Distribution and activity of nitrifying bacteria at two stations in the Ems estuary. Arch Hydrob Spe Iss Adv Limnol 47:65-76.

Guardabassi L, Gravesen J, Lund C, Bagge L, Dalsgaard A (2002) Delayed incubation as an alternative method to sample storage for enumeration of

E.coli and culturable bacteria in water. Water Res 36:4655-4658.

Hall GH (1982) Apparent and measured rates of nitrification in the hypolimnion of a mesotrophic lake. Appl Environ Microbiol 43:542-54

Illmer P, Schinner F (1995) Solubilization of inorganic calcium phosphates-solubilization mechanisms. Soil Biol Biochem 27:257-263.

Kim LH, Choi E, Stenstrom MK (2003) Sediment characteristics, phosphorus types and phosphorus release rates between river and lake sediments. Chemosphere 50:53-61.

Kim YH, Bae B, Choung YK (2005) Optimization of biological phosphorus removal from contaminated sediments with phosphate-solubilizing microorganisms. J Biosci Bio Eng 99:23-29.

Kittiwanich J, Yamamoto T, Kawaguchi O, Hashimoto T (2007) Analyses of phosphorus and nitrogen cyclings in the estuarine ecosystem of Hiroshima Bay by a pelagic and benthic coupled model. Estuar Coastal Shelf Sci 75:189-204.

Koops HP, Moller UC (1992) The lithotrophic ammonia-oxidizing bacteria. 2625-2637.

Lesniewska M, Witak M (2011) Diatoms as indicators of eutrophication in the SW part of the Gulf of Gdansk, the Baltic Sea. Oceanol Hydrobiol Stud 40(1):68-81.

Ling TY, Achberger EC, Drapcho CM, Bengtson RL (2002) Quantifying adsorption of an indicator bacteria in a soil-water system. Trans Am Soc Agric Eng 45:669-674.

Lipponen MT, Suutari MH, Martikainen PJ (2002) Occurrence of nitrifying bacteria and nitrification in Finnish drinking water distribution systems. Water Res 36:4319-4329.

Liu KK, Kao SJ, Wen LS, Chen KL (2007) Carbon and nitrogen isotopic compositions of particulate organic matter and biogeochemical processes in the eutrophic Danshuei Estuary in northern Taiwan. Sci Total Environ 382:103-120.

Lodi S, et al. (2011) Zooplankton community metrics as indicators of eutrophication in urban lakes. Nat Conserv 9(1):87-92.

Magalhaes CM, Joye SB, Moreira RM, Wiebe WJ, Bordalo AA (2005) Effect of salinity and inorganic nitrogen concentrations on nitrification and denitrification rates in intertidal sediments and rocky biofilms of the Douro River estuary, Portugal. Water Res 39:1783-1794.

Marina IS, Wm Brian A, Zhiyao S (2000) Oxygen uptake rate inhibition with PACT sludge. J Hazar Mater B 73:129-142.

Massa S, Caruso M, Trovatelli F, Tosques M (1998) Comparison of plate count agar and R2A medium for enumeration of heterotrophic bacteria in natural mineral water. World J Microbiol Biotechnol 14:727-730.

McCutcheon S (1987) Laboratory and in-stream nitrification rates for selected streams. J Environ Eng 113:628-646.

Mhamdi BA, Azzouzi A, Elloumi J (2007) Exchange potentials of phosphorus between sediments and water coupled to alkaline phosphatase activity and environmental factors in an oligo-mesotrophic reservoir. C R Biol 330:419-428.

Muirhead RW, Collins RP, Bremer PJ (2006) Interaction of Escherichia coli and soil particles in runoff. Appl Environ Microbiol 72:3406-3411.

Ng AS, Stenstrom MK (1987) Nitrification in powdered activated sludge process. J Environ Eng 113:1285-1301.

Oliver D, Clegg C, Heathwaite A, Haygarth P (2007) Preferential Attachment of Escherichia coli to Different Particle Size Fractions of an Agricultural Grassland Soil. Water, Air, Soil Pollut 185:369-375.

Palmer J, Flint S, Brooks J (2007) Bacterial cell attachment, the beginning of a biofilm. J Ind Microbiol Biotechnol 34:577-588.

Park SJ, Oh JW, Yoon TI (2003) The role of powdered zeolite and activated carbon carriers on nitrification in activated sludge with inhibitory materials. Process Biochem 39:211-219.

Pauer JJ, Auer MT (2000) Nitrification in the water column and sediment of a hypereutrophic lake and adjoining river system. Water Res 34:1247-1254.

Reyes I, Bernier L, Simard RR, Antoun H (1999) Effect of nitrogen source on the solubilization of different inorganic phosphates by an isolate of penicillium rugulosum and two UV-induced mutants. FEMS Microbiol Ecol 28:281-290.

Ruban V, López-Sánchez JF, Pardo P (2001) Harmonized protocol and certified reference material for the determination of extractable contents of phosphorus in freshwater sediments: A synthesis of recent works. Fresenius J Anal Chem 370:224-228.

Sahu SN, Jana BB (2000) Enhancement of the fertilizer value of rock phosphate engineered through phosphate-solubilizing bacteria. Ecol Eng 15:27-39.

Satoh K, Yanagida T, Isobe K, Tomiyama H, Takahashi R, Iwano H, Tokuyama T (2003) Effect of root exudates on growth of newly isolated nitrifying bacteria from barley rhizoplane. Soil Sci Plant Nutr 49:757-762.

Scott JA, Abumoghli I (1995) Modelling nitrification in the river Zarka of Jordan. Water Res 29:1121-1127.

Siemens J, Haas M, Kaupenjohann M (2003) Dissolved organic matter induced denitrification in sub-soils and aquifers? Geoderma 113:253-271.

Simon NS, Kennedy MM (1987) The distribution of nitrogen species and adsorption of ammonium in sediments from the tidal Potomac River and estuary. Estuar Coastal Shelf Sci 25:11-26.

Stief P, Schramm A, Altmann D, Beer D (2003) Temporal variation of nitrification rates in experimental freshwater sediments enriched with ammonia or nitrite. FEMS Microbiol Ecol 46:63-71.

Takahashi R, Ohishi M, Ohshima M, Saitoh M, Omata K, Tokuyama T (2001) Characteristics of an ammonia-oxidizing bacterium with a plasmid isolated from alkaline soils and its phylogenetic relationship. J Biosci Bioeng 92:232-236.

Thomsen L, Van Weering T, Gust G (2002) Processes in the benthic boundary layer at the Iberian continental margin and their implication for carbon mineralization. Prog Oceanogr 52:315-329.

Van MD, Portielje R, De NW, Boers PC (1998) Nitrogen in Dutch freshwater lakes: Trends and targets. Environ Pollut 102:553-557.

Vigdis T, Jostein G, Fride LD (1990) High diversity in DNA of soil bacteria. Appl

Environ Microbiol 56:782-787.
Wang BD (2006) Cultural eutrophication in the Changjiang (Yangtze River) plume: History and perspective. Estuar Coastal Shelf Sci 69:471-477.
Wang HY, Shen ZY, Guo XJ, Niu JF, Kang B (2010) Ammonia adsorption and nitritation in sediments resourced from Three Gorges Reservoir. China Environ Geol 60:1653-1660.
Wang LL, Niu JF, Yang ZF, Shen ZY, Wang JY (2008) Effects of carbonate and organic matter on sorption and desorption behavior of polycyclic aromatic hydrocarbons in the sediments from Yangtze River. J Hazard Mater 154:811-817.
White CS, Gosz JR (1987) Factors controlling nitrogen mineralization and nitrification in forest ecosystems in New Mexico. Biol Fert Soil 5:195-202.
Wu GF, Zhou XP (2005) Characterization of phosphorus-releasing bacteria in a small eutrophic shallow lake, Eastern China. Water Res 39:4623-4632.
Xu D, et al. (2011) Evaluation of the potential role of the macroalga Laminaria japonica for alleviating coastal eutrophication. Biores Technol 102(21):9912-9918.
Yang S, Zhao Q, Belkin IM (2002) Temporal variation in the sediment load of the Yangtze river and the influences of human activities. J Hydrol 263(1-4):56-71.
Yurkova N, Rathgeber C, Swiderski J, Stackebrandt E, Beatty LT (2002) Diversity, distribution and physiology of the aerobic phototrophic bacteria in the mixolimnion of a meromictic lake. FEMS Microbiol Ecol 40:191-204.
Zhang J, Liu SM, Ren JL, Wu Y, Zhang GL (2007a) Nutrient gradients from the eutrophic Changjiang (Yangtze River) Estuary to the oligotrophic Kuroshio waters and re-evaluation of budgets for the East China Sea Shelf. Prog Oceanogr 74:449-478.
Zhang J, Wu Y, Jennerjahn TC, Ittekkot V, He Q (2007b) Distribution of organic matter in the Changjiang (Yangtze River) Estuary and their stable carbon and nitrogen isotopic ratios: Implications for source discrimination and sedimentary dynamics. Mar Chem 106:111-126.

5

Chemical Effects

5.1 Overview

In recent years a worldwide increase in inputs of nitrogen (N) and phosphorus (P) into the lakes and reservoirs has led to considerable eutrophication and an increase in the frequency of toxic algae blooms (Carpenter et al., 1998; Cloern, 2001; Zeng et al., 2006; Fedro et al., 2007). The considerable changes in trophic composition and ecosystem structure (e.g. the plankton community) of the water environments adjacent to the lakes and reservoirs plumes have been attributed to the changes in Si:N:P ratios caused by excess of N and P, meaning cultural eutrophication (von Sperling et al., 2008; Burkholder et al., 2007; Yunev et al., 2007). For example, there are more than 30 lakes in southwest China. Although some of these lakes are seldom polluted, most of them have been heavily contaminated by recent human activities and they are experiencing a major deterioration of the water quality and a decrease in aquatic species. The increase in nutrient levels is believed to be an important factor in these problems.

Reservoirs and lakes are the significant accumulation body of fresh water, which has a close correlation with human activities. It is not only one of the most important drinking water resources, but also has functions of river runoff adjustment, flood prevention and disaster reduction, agricultural irrigation, breeding of marine products as well as electric power generation and tourism.

In China, there are various kinds of lakes and reservoirs. With the development of industrial production, water eutrophication will be increasingly serious if no measures are taken against it. In order to protect the water quality of the reservoirs and lakes, the status of water eutrophication must be effectively monitored, evaluated and predicted. Water eutrophication of reservoirs and lakes means that excessive nitrogen and phosphorus and other nutrient substances in the waters cause abnormal breeding of phycophyta and other hydrobios, which lead to changes in water transparency and dissolved oxygen (DO). Consequently, the deteriorating quality of water in reservoirs accelerates the damage to the ecosystem and water functions, which endangers the utilization of water resources (Wang et al., 2004).

Nitrogen and phosphorus overloading is a major cause of lake and reservoir eutrophication. The concentrations, transport and bioavailability of nitrogen and phosphorus in the lakes and reservoirs have been widely studied for many years by many researchers throughout the world. As known, excess nitrogen and phosphorus concentrations in water in lakes and reservoirs may cause eutrophication. When external loading of nutrients is increased, the sediments as a pool can adsorb them. However, after external loading is reduced, the sediments as a source will release adsorbed nutrients into the water. Many effects, such as the characteristics of sediments and overlying water quality, will affect the transfer direction of nutrients at the sediment–water interface. Therefore, intensive study of the main influencing factors for the transport of nutrients would help to predict and further prevent the occurrence of eutrophication.

Nitrogen is a ubiquitous biogenic element and widely distributed in the nature. The discharge of nitrogen into the receiving waters has a substantial influence on water quality and the health of fish and other aquatic organisms. Reduced nitrogen forms can be oxidized in fresh water systems causing oxygen depletion (Sharma and Ahlert, 1977) which can seriously impair or even kill aquatic organisms (e.g. Wetzel, 1983; Thomann and Mueller, 1987). Nitrogen is also one of the major nutrients controlling phytoplankton growth and standing crops (Welch, 1992; Wetzel, 1983).

In aquatic systems the main forms of nitrogen are ammonium-nitrogen (NH_4^+-N), nitrite-nitrogen (NO_2^--N), nitrate-nitrogen (NO_3^--N), organic nitrogen and nitrogen gas (N_2). The dominating form is determined by the environmental factors of the water body such as pH, temperature, oxygen and microorganism activity coupled with the mineralization rates of organic nitrogen. Seasonal changes can also be a key control of the speciation balance regardless of the total nitrogen concentration of the water body (Burt et al., 1993). Excess nitrate can contribute to eutrophication and can also be toxic for some aquatic organisms.

Different species of nitrogen reciprocally transform in water. This process is influenced by many factors including physical, chemical, biological and hydrodynamic effects. In the other chapters in this book, biological and hydrodynamic effects have been introduced. So in this chapter, physical and chemical factors will be emphasized.

Temperature can directly affect the activity of microorganism enzymes. So, in a certain range, the higher the temperature is, the more the transformation of the elements. Nitrification is an exoergic process which involves O_2 and microbial organisms, and heat dissipation contributes to the nitrification process (Wang et al., 2004). The optimum temperature of nitrifying bacteria appears to be within the range of 25–35 °C, and the temperature range in which bacteria growth happens is usually 3–45 °C (Focht and Verstraete, 1977). The optimum temperature for natural population growth depends largely on ambient temperature. For example, Jones and Hood (1980a; 1980b) reported that the optimum temperature was 40 °C for the growth of nitrosomonas species isolated from a subtropical estuarine bay in Florida while, for the same species isolated from the temperate Ems-Dollard estuary in Holland, Helder and De Vries (1983) observed that the optimum temperature was 25–35 °C.

Temperature can inhibit or accelerate the release speed of different nitrogen species from sediment by changing microbe activities. Many studies have demonstrated a similar result for the rate of nitrogen released by lake sediments. That is, a higher nitrogen release rate was observed in summer and less in winter. For example, Taihu Lake sediment is the source of the release of nitrogen in summer (25 °C), while nitrogen in overlying water will be transfered to the sediment in winter (5 °C). In winter the sediment is the sink of nitrogen (Fan et al., 2004). The reason may be the microbial effect. When the temperature is high, the microbes in the sediment are active, which can speed up the decomposition of organic nitrogen or mineralization of organic matter. This would accelerate the release of nitrogen from sediment to water.

DO is an important factor influencing nitrogen release. Different nitrogen species at different DO levels present a different transfer rule. Nitrous oxide is produced within the first step by microbes in the nitrification process at low oxygen concentrations (Goreau et al., 1980; Hynes and Knowles, 1984). In the measurements of nitrification and N_2O production, N_2O production increased sharply when oxygen was below 0.1811 mol/L and showed a maximum in complete anoxic conditions. In pure cultures of a marine nitrosomonas species, the production of N_2O per cell increased at the lowest oxygen level (5.611 mol/L O_2), whereas the production of NO_2^- per cell declined (Goreau et al., 1980).

Some researches showed that the nitrogen release rates from the sediment increased with the reduction in dissolved oxygen and ammonia nitrogen was the dominant species in the dissolution of inorganic nitrogen in anaerobic conditions, while nitrate nitrogen dominated in aerobic conditions. For example, increasing inhibition of potential nitrification activity at increasing O_2 levels has been measured in an estuarine sediment. A 15% and 25% inhibition to the nitrification process have been observed at 2 and 2.6 times air saturation respectively (Revsbech et al., 1981). Bai et al. (2002) reported that nitrate nitrogen and nitrite nitrogen released from urban river sediment were near to zero when the DO concentration was less than 1.5 mg/L, while the ammonia nitrogen release speed could reach 10 mg/(m^2·day). Furthermore, the ammonia nitrogen release quantity fell sharply when DO concentration was more than 2.8 mg/L. Fan and Morihiro (1997) also observed that, in anaerobic conditions, the ammonia nitrogen release rates from the sediment of Tsuchiura Bay were 2–8 times as much as in aerobic conditions and the released total inorganic nitrogen observed from sediment was higher than in aerobic conditions. Li et al. (2003) found that, in aerobic conditions, most ammonia nitrogen can be transformed into nitrate by nitrifying bacteria through the nitrification process and then released into overlying and pore water. In this process, the concentration of ammonia nitrogen released from the sediment decreased sharply. With the decrease in dissolved oxygen, ammonia nitrogen became the key species in the released total nitrogen. Furthermore, organic nitrogen and total nitrogen increased but nitrate nitrogen and nitrite nitrogen reduction had also been observed. The release rate of nitrogen is low in aerobic conditions because the demand of nitrogen for organisms in aerobic conditions is very high. At this time, the nitrogen release from sediments is usually less than that in anaerobic conditions (Wu and Morihiro, 2005). Nowadays, more and more researchers consider that maintaining a proper dissolved oxygen level can effectively restrain the nitrogen nutrient release from sediments, especially the release of ammonia nitrogen (Fan and Morihiro, 1997; Jiang et al., 2007).

As an important influencing factor on nutrient release, the pH value has been widely studied. The form of nitrogen depends on the water pH value. For example, the NH_3-N existing in a water environment mainly refers to the free ammonia (NH_3) and ammonium ion (NH_4^+). If the pH value is high, the proportion of free NH_3 is on the high side.

The activity of nitrifying bacteria is optimal in a narrow pH range from neutral to slightly alkaline (pH 7–8.5), with the limiting range for activity being slightly wider (pH 6–9.5) (Focht and Verstraete, 1977). The upper limit is dependent on the concentration of free (undissociated) ammonia (NH_3), which is toxic to the bacteria. On the other hand, it has been suggested that undissociated NH_3 rather than NH_4^+ is the substrate for ammonium oxidizers (Suzuki et al., 1974), such that slightly alkaline conditions would increase substrate availability, which would be beneficial at low NH_4^+concentrations in surface sediment. Optimum rates at pH 7.5–8.5 were measured for an estuarine nitrosomonas sp., whereas the activity at pH 9 was still 50% of maximum (Jones and Hood, 1980b). It is generally acknowledged that pH can affect the microbes in sediments, further influencing nitrogen transformation in the interface of the water-sediment. In a strong acidic environment, pH could suppress the activity of nitrifying bacteria, denitrifying bacteria, and microbial ammonia oxidation. The release quantity was not apparent, when pH went to slightly alkaline conditions. Ammonia oxidizing bacteria were more active and promoted the release of nitrogen in sediments when pH continued to increase, and the release quantity decreased as the microbial activity reduced.

Many studies have found that the effect of pH on the release of nitrogen from sediments is obvious. A study showed that no change happened in the process of nitrogen release from sediment when pH was within the range of 3–6. The nitrogen release quantity decreased when $pH < 3$ and was the largest when $6 < pH < 8$. As pH further increased, the nitrogen release quantity reduced sharply and the release quantity of ammonia nitrogen even dropped 70% (Liu et al., 2002). Generally, there occurred only small pH fluctuations in marine sediments in deeper waters, whereas surface sediments in shallow-water with benthic microalgae can show dramatic diurnal variations in pH (e.g. pH 7–10) (Revsbech et al., 1981; Andersen and Jensen, 1983). In fact, acidity and alkalinity to some extent decide the oxidation and reduction environment. Redox potentials (E) and pH value have negative correlation because electronic and H^+ have the opposite charge (Zhao and Yang, 2009). In an aerobic situation, higher redox potentials exist in the water-sediment interface. At this time an electron may be changed from a reduction state to an oxidation state. From the electrochemical aspect, the lower the pH value is, the higher the concentration of H^+ ions in the solution. NH_4^+ in the sediment colloids can have competitive adsorption with H^+ ions and be released. The mechanism of the pH influence is complex. Generally, researchers only do the study of the pH influence on total nitrogen and ammonia nitrogen while other different forms of nitrogen release have been less considered. As the pH of these sediments and overlying water vary, it could be an important factor regulating the nitrification rate in the sediments.

In the nitrification process, nitrifying bacteria are susceptible to inhibition because of their slow intrinsic growth rate and their sensitivity to a number of environmental conditions and other ions. Consequently, nitrification inhibition by metals has attracted considerable attention (Hu et al., 2003).

Transport of metal across the bacterial cell membrane has been studied extensively. Sorption plays an important role in metal uptake (Battistoni et al., 1993) because extracellular polymers and bacterial cell surfaces contain a variety of binding sites including amino, carboxylic, hydroxyl and phosphate functional groups because of their (phospho)lipid, protein and polysaccharide moieties (Hu et al., 2003). There is little research about the effect of heavy metal on nitrogen in Chinese rivers. Many researchers have done experiments on the effects of heavy metals on bacteria. Yang et al. (2008) showed that the nitrate reducing process was likely to occur if there were high levels of manganese (Mn) or manganese and iron (Fe, Mn) in water samples of the Hunhe River, which could make NO_3^--N turn to NH_4^+-N. The sample with a high concentration of Fe has a weak effect on the nitrate reduction while low levels of Fe in water samples could improve the nitrification rate. Low levels of Mn in water samples almost have no effect on nitrification. Then we learnt that the low levels of Fe can act as a role in promoting the nitrification process, while a high concentration of Fe possesses inhibition to nitration; The existence of Mn and the coexistence of Fe and Mn both have an inhibition effect on the nitrification process. Juliastuti et al. (2003) reported that zinc and copper inhibited the nitrite and nitrate production at different concentrations. Copper had a stronger inhibitory effect than zinc: Zn^{2+} inhibited nitrification at a concentration exceeding 0.35 mg/L, whereas Cu^{2+} already inhibited at a concentration above 0.05 mg/L. Another study showed that heavy metal ions (Cu^{2+}, Cd^{2+} and Zn^{2+}) in the sea water of Dalian Bay had more inhibition effects on nitrosation than on the nitrification process. The efficiency of the two processes was inhibited as follows: $Cu^{2+}>Cd^{2+}>Zn^{2+}$, and the inhibition degree increased with the increasing concentration of the ion. In addition, the nitrification process was difficult to recover if Cu^{2+} had inhibition in it (Weng et al., 2006). Furthermore, Hu et al. (2003) observed that the inhibitory and sorption mechanism of Cu^{2+} was very different from Zn^{2+}, Ni^{2+}, and Cd^{2+} and might involve rapid loss of microbial membrane integrity. The inhibition behavior of the same heavy metal in other circumstances might be different, so more research is required in the future.

It is widely accepted that sediments are important sites for the mineralisation of organic matter (Jorgensen and Revsbech, 1989). Ammonium is produced in sediments during the decomposition of organic matter by various heterotrophic organisms. After its release to porewater solution, NH_4^+ can be reoxidized, reincorporated into organisms, adsorbed onto particles, or can diffuse along concentration gradients to other regions of the sediment (Martens et al., 1978). These all will have great effects on the nitration and denitrification process in the sediments. Therefore, the adsorption and desorption of ammonia nitrogen on the sediments play an important role in nitrogen cycling. The physical and chemical properties of sediments have a significant influence on the behavior of ammonia nitrogen adsorption in different depositional environments.

Mackin et al. (1984) reported that there occurred a positive correlation of sediment porosities with the adsorption coefficient of ammonia nitrogen in western Atlantic sediments. In very high porosity (≥0.95) sediments, adsorption coefficient may be somewhat lower than for other sedimentary environments. The author considered that at high sediment porosities, the grain surface area became limiting. In the study of sediments of shallow lakes in the middle reaches of the Yangtze River area (Jiao et al., 2007), the highest release rate of ammonia nitrogen had no significant correlation with CEC, TOC and Fe/Al hydroxides but had good agreement with the content of the fine grain size fraction. The result showed that the relative contribution of the fine grain size fraction to nitrogen cycling was the absolute main fraction, which was several to 10 times that in the coarse grain size fraction. While another study (Boatman and Murray, 1982) has compared the ammonium adsorption constants before and after the extraction of organic matter, the result indicated that in organic-rich sediments the organic matter or a "clay-humic complex" might be controlling ammonium adsorption. However, in organic-poor sediments, the clay mineralogy will tend to dictate ammonium adsorption behavior. Another research in the Yangtze estuary demonstrated that sediment porosities and the content of fine particulate solids (<63 μm) didn't have a significant correlation with ammonium adsorption (a>0.05). To the contrary, there existed good agreement between TOC content and ammonium adsorption (a>0.05), indicating that the organic matter might control the ammonia nitrogen adsorption behavior in the sediments of the Yangtze estuary (Hou et al., 2003). In conclusion, the physical and chemical properties of sediments play an important role in the transformation of nitrogen in sediments

Phosphorus overloading is the most important cause of lakes and reservoirs eutrophication (Wang et al., 2007). As a consequence of human activity, phosphorus arrives in natural aquatic ecosystems and accumulates in sediments. Under certain environmental conditions, phosphorus can be released from the sediments to the overlying water. The transport and transformation of P in water environments involve a great number of complex processes, both biotic and abiotic. The phosphate sorption-desorption at the sediment–water interface is an important process affecting the phosphorus (P) transport, bioavailability and concentrations in lakes and reservoirs. During periods of enhanced external loading the sediments can take up phosphate, but after reduction of external inputs the sediments can release adsorbed phosphate and thus become a source of eutrophication. Therefore, the sediment–water exchange of phosphate becomes more important for eutrophication in the lakes and reservoirs and P has been identified as a limiting nutrient to the biological production.

The phosphate sorption-desorption capacity at the sediment–water interface varied with sediments components, particles size, solution parameters and so on (Kim et al., 2003; Gardolinski et al., 2004). All factors have to be taken into account when evaluating phosphate sorption-desorption by lake and reservoirs sediments. Many similar results reported that many sediment properties, e.g. particle size, organic matter, Fe- and Al-oxides and cationic exchange capacity (CEC), affected the phosphate sorption-desorption process. Organic matter plays an important role in sediments since its mineralization may result in changes in both redox and pH and humic substances can stabilize iron particles. Sediments where an important fraction of phosphate occurs adsorbed at the surface of metal oxides or other minerals usually show a relatively slow release in near neutral or weakly acidic media and a relatively fast release in alkaline media. Ionic strength is also a factor affecting phosphate sorption at the sediment–water interface. The high ionic strength in solution lowered the extent of sorption and shifted the sorption isotherms toward higher equilibrium phosphate concentrations. pH in water is another important factor (Ku et al., 1978; Lijklema, 1980; Crosby et al., 1984; Seitzinger, 1991). The fraction of phosphate adsorbed at the surface of metal oxides or other minerals in the sediment usually shows a relatively slow release in near neutral or weakly acidic media and a relatively fast release in alkaline media. These effects are attributed to the decreasing affinity that mineral surfaces, especially those of Fe and Al oxides, show for phosphate when the pH is increased, and to competition between hydroxyl and phosphate ions for surface sites. On the other hand, sediments where an important fraction of phosphate occurs as calcium phosphate minerals may also show an important and fast phosphate release at low pH, where these minerals preferentially dissolve.

Another factor affecting the rate of phosphate release is temperature. There is some experimental evidence indicating that the rate increased by increasing the temperature (Kim et al., 2003). This effect is easily understood in terms of Arrhenius-type expressions, since an activation energy (E_a) barrier needs to be overcome during the release process. The magnitude of E_a, besides providing the temperature dependence of the reaction rate, often gives relevant information regarding the mechanism of the process (Borgnino et al., 2006).

ORP is one of the most important parameters to describe the phosphorus adsorption onto the sediments. Ironbound-phosphorus is particularly sensitive to redox and pH variations of the system. The very low ORP induces the reduction of Fe(III) from the Fe(OOH) in the sediment. Zhou et al. (2005) reported that ORP decreased at different rates during the pH range observed. When pH is very low, ORP is very high, which should, in principle, produce high phosphorus adsorption, but the amount of P sorbed decreases rapidly in this range of pH. Therefore, in this kind of condition, although high affinity to adsorb phosphorus caused by high ORP of particles is produced, H^+ still overcomes particles to trap SRP because of its very high concentration as was interpreted in the zeta potential effect. As pH further increases, the positive correlation between P sorbed and ORP can be approximately observed, but they don not vary exactly in phase, which indicates that ORP will significantly influence P adsorption but cannot completely control the overall adsorption process. With the further increase in pH, the reduction of Fe (III) and the competition from OH^- mentioned above would both hinder phosphorus adsorption onto sediment.

Besides all the mentioned factors, surface properties of sediment particles, such as surface charge density and specific surface area, can also be important factors for the rate of phosphate release. This is especially true for sediments where an important fraction of the total phosphate occurs at the surface of mineral particles. Changes in surface charge may be mentioned in some articles as responsible for the changes in the release rates, but little direct measurement of this charge has been reported. Therefore, relative study is still needed in order to evaluate the effects of surface charge on the release rate.

The Three Gorges Dam, in China, is the world's largest dam, measuring 2335 m long and 185 m high, and the reservoir created by it had an area of 1080 km^2 in 2009 (Wu et al., 2003). The construction of the Three Gorges Reservoir will have a profound impact on the environment (Jiao et al., 2007), wherein eutrophication would be a potential threat (Zeng et al., 2006). Phosphorus is a major nutrient for aquatic ecology, and its excess supply can lead to eutrophication. Especially in most lakes and reservoirs in China, phosphorus is the main limiting factor for eutrophication. Therefore, the chemical effects of phosphorus transfer were the focus in this chapter.

In this chapter, the adsorption capacity for phosphorus on the sediments from Cuntan, Xiaojiang, Daning and Xiangxi in the Three Gorges Reservoir, China, was evaluated. Phosphate sorption-desorption capacity at the sediment–water interface as a function of sediment components, pH, temperature, ionic strength, and so on, were discussed systematically. The aim was to provide information on the effect and mechanism of the sediments characteristics and overlying water quality on the nutrients transport.

5.2 Sediment Components

5.2.1 Sampling Locations and Properties

Sediment samples were collected at the influx of four main branches (Cuntan, Xiaojiang, Daning and Xiangxi River) of the Three Gorges Reservoir on the Yangtze River using a Van Veen stainless steel grab sampler (Eijkelamp, Netherlands) in October 2006 (Fig. 2.1). The sampling site is subject to multiple pollution sources including shipping activities, urban runoff and combined sewer overflows from ambient cities. The samples were taken to the laboratory in air-sealed plastic bags and were kept at 4 °C. They were then air-dried at room temperature and ground to pass through a 100-mesh sieve for adsorption experiment. Many studies have demonstrated that adsorption capacity for P on different sediments was related to sediment composition, such as contents of organic matter, Fe/Al hydroxides, clay and $CaCO_3$ (Nwoke et al., 2003; Makris et al., 2005; Lake et al., 2007). Wang et al. (2007) reported that there occurred a positive correlation of organic matter with P adsorption. Sanyal et al. (1991) found that the content of clay had a significant correlation with P adsorption. The content of metals was thought to be the main factor that determined adsorption capacity, because of the high specific surface of the Fe/Al hydroxides. Additionally, the regulation of dissolved phosphate by adsorption or coprecipitation with calcite formed in situ has been widely reported in calcareous systems (Berg et al., 2004).

That is, solid phase $CaCO_3$ usually governs P reaction in calcareous soil (Berg et al., 2004). Therefore, the P adsorption process by sediments has been affected obviously by sediment composition.

At present there are many ways to study the effects of solid composition on adsorption, in which a sequential chemical extraction is regarded as a powerful method and consequently has been widely used for soil adsorption study for many years. In 1979, this method was developed by Tessier et al. (1979) who successfully used it in discussing the trace metal speciation on the sediments. Subsequently, Tang et al. (1982) utilized an improved extraction method to study the adsorption capacity for metal in sediments.

In this section, the surface properties of sediments from the Three Gorges Reservoir were investigated. The relation between the adsorption capacity for phosphorus in the sediments and the sediments composition was further analyzed. What's more, the effect of sediment composition on P adsorption by a sequential chemical extraction method was conducted.

5.2.2 Sediments Characteristics

Generally, the retention capacity of the sediments for phosphorus was obviously influenced by their characteristics such as contents of organic matter, metal hydroxides (Fe, Al), $CaCO_3$ and clay (Nwoke et al., 2003; Makris et al., 2005; Lake et al., 2007). In this study, the main characteristics of sediments from the influx of four main branches (Cuntan, Xiaojiang, Daning and Xiangxi River) of the Three Gorges Reservoir on the Yangtze River were summarized in a section of Chapter 2 in this book.

5.2.3 Adsorption Capacity of Different Sediments for Phospate

Adsorption kinetics are important as they can provide valuable insights into the mechanism of sorption reactions. To clarify the mechanism of P adsorption, much of the emphasis has been focused on developing mathematical models to describe P adsorption characteristics on natural sediments. Lopez et al. (1996) reported that adsorption kinetics of phosphate on the sediments followed a nonlinear model. Tu et al. (2002) adopted Langmuir, Freundlich and Temkin equations to fit the P adsorption isotherm, and found that the Langmuir model was the best. This result was also obtained by many other researchers (Lopez et al., 1996; House and Denison, 2000; Bubba et al., 2003). However, when the existence of native adsorbed P (NAP) was concerned, adopting directly the Langmuir model was unreasonable because NAP in sediment also took part in the adsorption equilibrium. To improve it, Zhou et al. (2005) brought forward a modified Langmuir model which could describe well P adsorption on sediments with NAP.

Fig. 5.1 shows the adsorption kinetics of phosphate on different sediments. It

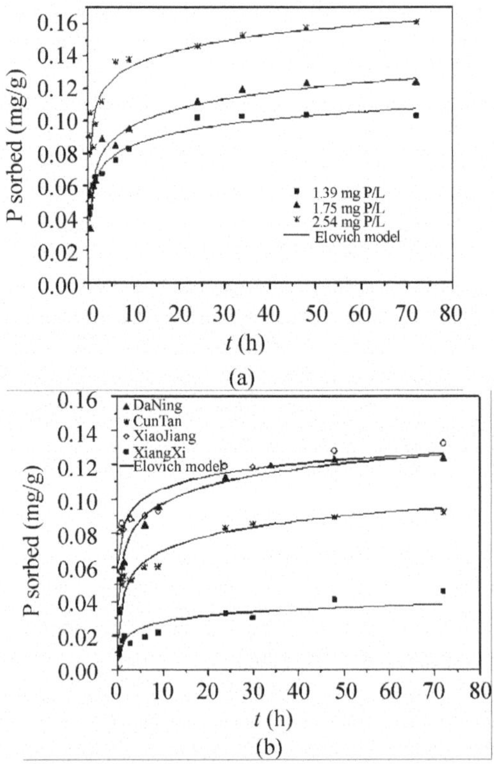

Fig. 5.1 Adsorption kinetics of phosphate (a) on the Daning sediment at different initial P concentrations; (b) on the different sediments at the initial P concentration of 1.75 mg P/L

can be seen that phosphate adsorption on the sediment included quick and slow adsorption steps. The quick adsorption step mainly occurs within 6 h, and is then followed by a slower second step. After 48 h, the adsorption amount of phosphate levels off, indicating that a pseudo-equilibrium occurred. The adsorption amount and the rate of the initial stage for phosphate adsorption advanced with an increase in the initial phosphate concentration. The similar phenomenon is also observed in other sediments (Fig. 5.1(b)). At the same initial P concentration, the adsorption capacities for various sediments followed the order: Xiaojiang>Daning>Cuntan> Xiangxi.

The adsorption kinetics analysis was tested using two kinetic models, namely power function and the Simple Elovich model. The estimated parameters of kinetic models for all the sediments studied here are summarized in Table 5.1. Based on R^2 values, the power function and Simple Elovich model can both best describe adsorption kinetics of phosphate and there is no clear difference between them.

Table 5.1 Kinetic model parameters of phosphate adsorption on the different sediments from the influx of four main branches (Cuntan, Xiaojiang, Daning and Xiangxi River) of Three Gorges Reservoir on the Yangtze River

Samples	Initial concentration (mg P/L)	Power function model ($Q=a\times t^b$)			Simple Elovich model ($Q=a+b\ln t$)		
		a	*B*	R^2	*a*	*b*	R^2
Cuntan	1.39	0.034	0.242	0.93	0.035	0.013	0.95
	1.75	0.040	0.209	0.87	0.041	0.013	0.92
Xiaojiang	1.39	0.066	0.131	0.94	0.068	0.01	0.91
	1.75	0.080	0.115	0.91	0.081	0.011	0.89
Xiangxi	1.39	0.010	0.276	0.95	0.011	0.004	0.88
	1.75	0.014	0.274	0.94	0.015	0.005	0.85
Daning	1.39	0.058	0.153	0.96	0.059	0.011	0.97
	1.75	0.062	0.175	0.93	0.064	0.015	0.93
	2.54	0.101	0.116	0.92	0.102	0.014	0.91
Daning A1	1.39	0.057	0.094	0.97	0.057	0.006	0.98
	1.75	0.060	0.107	0.98	0.061	0.008	0.96
	2.54	0.078	0.116	0.90	0.079	0.011	0.91
Daning A2	1.39	0.027	0.104	0.77	0.027	0.003	0.79
	1.75	0.031	0.106	0.93	0.031	0.004	0.92
	2.54	0.038	0.117	0.92	0.039	0.005	0.89

Adsorption capacity at different phosphate concentrations can be illustrated by the adsorption isotherm (Wang et al., 2009). As shown in Fig. 5.2 (Wang et al., 2009), higher phosphate concentrations in water increase the adsorption amount of P on the sediments. When the P concentrations in water were low, phosphate adsorbed on the sediments was released. However, when P concentrations in water were higher, phosphate in water was adsorbed by the sediments. That is, P will be exchanged on the sediment–water interface until a dynamic equilibrium is reached. This process can be analyzed by the adsorption isotherm model.

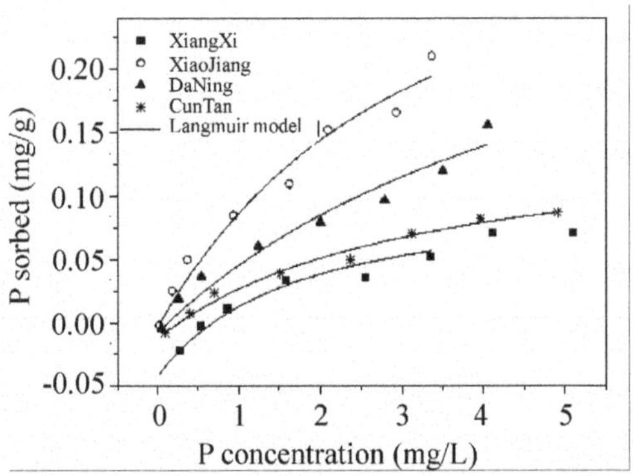

Fig. 5.2 Adsorption isotherms of phosphate on different sediments from the influx of four main branches (Cuntan, Xiaojiang, Daning and Xiangxi River) of the Three Gorges Reservoir on the Yangtze River at 20 °C

When sediment is in contact with water, P will be exchanged with the water on the interface until a dynamic equilibrium is reached. The equilibrium situation is usually described by a Langmuir adsorption isotherm.

$$q_e = Q_{max} \cdot P_{eq} / (k + P_{eq}) \tag{5.1}$$

where q_e is the mass of net adsorbed P per mass unit at equilibrium (mg/g), P_{eq} is the equilibrium concentration of P in solution (mg/L), Q_{max} is the maximum P adsorption capacity at saturation conditions per mass unit of sediment (mg/g), k is the empirical constant with unit of concentration (mg/L) and corresponds to the concentration at which half the maximum sorption capacity is reached and is related to the surface binding energy. The Langmuir model assumes a homogeneous surface, uniform binding energy and monolayer coverage. Because the amount of total native (or original) exchangeable P (NAP) in sediment also takes part in the sorption equilibrium, the equilibrium conditions of adsorption trials can be described as (Zhou et al., 2005):

$$(NAP + C_{add}/w) - P_{eq}\ V/w = Q_{max} \cdot P_{eq} / (k + P_{eq}) \tag{5.2}$$

or

$$(C_{add} - P_{eq}) V/w = Q_{max} \cdot P_{eq} / (k + P_{eq}) - NAP \tag{5.3}$$

where NAP refers to the amount of total native (or original) exchangeable P (mg/g), w is the dry weight of sediment in the sorption trials (g), C_{add} is the initial concentration of newly added P in solution of adsorption trials (mg/L), V is solution volume (L). P_{eq}, Q_{max} and k has been described in Eq. (5.1), $P_{eq}V$ refers to the amount of exchangeable P remaining in solution of adsorption trials, which should include both NAP dissolved and newly added P that was not adsorbed by sediment. Considering an extreme condition with C_{add}=0; then Eq. (5.3) will take the form

$$-P^0_{eq} V/w = Q_{max} \cdot P^0_{eq} / (k + P^0_{eq}) - NAP \tag{5.4}$$

where P^0_{eq} denotes equilibrium concentration of exchangeable phosphorous in the solution with initial C_{add}=0; which actually is solution concentration in desorption equilibrium by fresh water. In this special case, there is no exchangeable P added and thus all of exchangeable P in the system is native exchangeable P. In Eq. (5.4), when desorption/adsorption processes fall into dynamic equilibrium, the amount of exchangeable P adsorbed is described by the value of the left part. Then *NAP* can be obtained by

$$NAP=Q_{max}\cdot P^0_{eq}/(k+P^0_{eq})+P^0_{eq}V/w \tag{5.5}$$

When *NAP* in Eq. (5.4) is substituted by Eq. (5.5), the resulting equation will hold

$$(C_{add}-P_{eq})V/w = Q_{max}\cdot P_{eq}/(k+P_{eq})-(Q_{max}\cdot P^0_{eq}/(k+P^0_{eq})+P^0_{eq}V/w) \tag{5.6}$$

Eq. (5.6) can serve as a model to describe P sorption properties on lake sediments that contain different amounts of *NAP*. Using this model, the data of sorption isotherm trials can be nonlinearly fitted properly with the least square method. The values of Q_{max} and k are then obtained. Afterwards, *NAP* can be calculated by Eq. (5.5) with corresponding parameters. EPC_0 refers to water phase concentration at which P_{eq}=C_{add} where the original sediment and water phosphate concentrations are in dynamic equilibrium. When P_{eq}=C_{add}=EPC_0; the expression of EPC_0 according to Eq. (5.6) will take the form below:

$$EPC_0 = P_{eq} = \frac{\dfrac{k\cdot Q_{max}\cdot P^0_{eq}}{k+P^0_{eq}} + \dfrac{k}{w}P^0_{eq}V}{\dfrac{k\cdot Q_{max}}{k+P^0_{eq}} - \dfrac{1}{w}P^0_{eq}V} \tag{5.7}$$

and K_p can be calculated by

$$K_p = \frac{NAP}{EPC_0} \tag{5.8}$$

Table 5.2 showed the modified Langmuir model parameters. Obviously, the modified Langmuir model had a good fit with the experimental data for the four sediments (R^2>0.96). Over the entire concentration range studied, the adsorption of phosphate by different sediments is in the order of Xiaojiang>Daning> Cuntan>Xiangxi. This sequence was the same as the result of a kinetics study. Theoretically, the maximum adsorption amount (Q_{max}) was 0.402 mg P/g for Xiaojiang sediment, 0.358 mg P/g for Daning sediment, 0.165 mg P/g for Cuntan sediment, and 0.15 mg P/g for Xiangxi sediment.

Adsorption capacity has been related with sediment characteristics. According to other reports (Brinkman et al., 1993), the content of metals seems to be the main factor that determines adsorption capacity, because of the high specific surface of the iron/aluminium hydroxides. In some cases, the adsorption sites can be occupied by phosphate, so the ratio of P/(Fe+Al) has been used to provide a better measure of the free sorption sites for phosphate in sediments. Moreover, for the calcareous sediments, a large fraction of the sedimentary phosphate associated with calcium minerals should be taken into account.

Table 5.2 Some adsorption parameters obtained from the modified Langmuir model for different sediments

Samples	Q_{max} (mg/g)	k (mg/L)	R^2	P^0_{eq} (mg/L)	P_w (mg/L)	NAP (mg/g)	EPC_0 (mg/L)	K_p
Cuntan	0.165	3.186	0.986	0.098	0.036	0.013	0.266	0.048
Xiaojiang	0.402	3.421	0.972	0.022	0.050	0.004	0.037	0.116
Xiangxi	0.150	1.713	0.967	0.272	0.082	0.042	0.674	0.063
Daning	0.358	5.743	0.959	0.054	0.033	0.008	0.126	0.061
Daning A1	0.205	3.921	0.991	0.040	-	0.005	0.103	0.051
Daning A2	0.083	3.433	0.970	0.022	-	0.002	0.098	0.023

As shown in Table 5.1 and Table 5.2, for the Xiangxi sediment, because of the lowest concentrations of (Fe+Al+Ca) and free adsorption sites (the highest ratio of P/(Fe+Al)), the capacity for phosphate adsorption was the lowest. Xiaojiang sediment presented the highest concentrations of organic matter and (Fe+Al), the lowest ratio of P/(Fe+Al). So the phosphate adsorption capacity of the Xiaojiang sediment was the highest, which was in accordance with the result of adsorption kinetics and isotherm. For Daning and Cuntan sediments, results were more difficult to interpret. Although the contents of (Fe+Al) and the free sorption sites for phosphate analyzed from the ratio P/(Fe+Al) were both higher in Cuntan than in Daning sediment, a higher adsorption capacity for phosphate was observed in Daning sediment. As reported, the calcareous matter also exhibited the adsorption or coprecipitation for phosphate (Berg et al., 2004). So higher adsorption capacity of Daning sediment may be related to higher concentrations of calcium and organic matter.

As shown in Table 5.3, *NAP* values, the intersection of *y*-axes and the regression curves, range from 0.004 mg/g in Xiaojiang to 0.042 mg/g in Xiangxi sediment. EPC_0, the *x*-intercept of the model, was defined as the phosphate equilibrium concentration in water, at which there was no net adsorption or release of phosphate on sediments. When EPC_0 was higher than actual phosphate concentration in water (P_w), the sediment would release P. Inversely, when EPC_0 was less, the sediment would adsorb P from water. That is, higher EPC_0 increased the risk of P releasing from the sediment to water. In our study, the highest EPC_0 values of 0.674 mg/L calculated by adsorption isotherm were obtained for the Xiangxi sediment. So the Xiangxi sediment showed the highest risk of P releasing from the sediment to water. Comparing EPC_0 with P_w values shown in Table 3, the values of P_w in all the sediments except for Xiaojiang were lower than that of EPC_0. This indicated that the sediments in Cuntan, Daning and Xiangxi have a trend of releasing P as a source role, while the Xiaojiang sediment might adsorb P from water as a pool role.

Table 5.3 The characteristics of each grade sediment extrated from the Daning sample

Samples	TOC (%)	Fe (mg/g)	Al (mg/g)	Ca (mg/g)	Clay (%)	Silt (%)	Sand (%)
Daning	0.51	23.63	7.75	101.32	45.89	53.02	1.10
Daning A1	0.51	21.01	6.33	69.62	33.14	62.54	4.32
Daning A2	0.24	12.19	3.83	53.41	60.36	39.63	0.006

K_p refers to the ratio of *NAP* and EPC_0, which denoted the attraction forces towards P. The higher value of K_p the sediment possessed, the stronger the attraction forces towards P. In our study, the maximum value of K_p was observed in Xiaojiang sediment. So Xiaojiang sediment may hold the highest attraction forces towards P. That is, compared with other sediments in our study, P in water was easier to be adsorbed by Xiaojiang sediment, which was in accordance with the analysis of EPC_0 above mentioned.

5.2.4 Effect of Sediment Compositions

Obviously, based on the above discussion, sediments compositions were considered to be the major factors controlling phosphate adsorption by sediments. Therefore, a sequential chemical extraction experiment described by Tang et al. (Tang et al., 1982) was designed to further clarify the effect of different composition. In this study, Daning sediment was taken as an example and divided into three fractions by sequential chemical extraction.

Detailed procedure was as followed.

1) Removal of carbonates and exchangeable calcium

The sediment was extracted at room temperature for 6 h by 1 mol/L NaOAc adjusted to pH 5.0 with acetic acid (HOAc) with continuous agitation. After stewing overnight, the sediment was filtered, washed thoroughly with deionised water and subsequently dried at 60 °C. This residual sediment was mentioned as A1.

2) Removal of organic matter and hydrated metal oxides

Hydrogen peroxide solution was added dropwise into a triangular flask containing the residue obtained from step 1) at 40 °C with intermittent agitation, and a large number of bubbles were generated. After 12 h reaction, the mixture was evaporated to dryness. To remove organic matter thoroughly, the above steps were repeated three times. Then the residue was extracted by mixed solution (0.8 mol/L $H_2C_2O_4$ and 1.2 mol/L $(NH_4)_2C_2O_4$, pH=3.2) three times in order to prevent adsorption of extracted phosphorus onto the oxidized sediment and remove metal hydroxides. This residual sediment was mentioned as A2.

Daning A1 denoted the residue which was obtained by depriving carbonates and exchangeable calcium partially from Daning sediment. After getting rid of organic matter and metal hydroxides from Daning A1, the residue, which was mainly composed of clay mineral, was mentioned as Daning A2.

The difference in sediment characteristics before and after treatment was shown in Table 5.3. It can be seen that no organic matter was removed for Daning A1 sediment, while 31.3% of Ca was reduced. Simultaneously, a small quantity of Fe and Al dissolved. For Daning A2 sediment, more than half of the organic matter was lost. About 50% of Fe, Al and 47.3% of Ca dissolved. Those results were consistent with the expected values. Additionally, the proportion of clay in the Daning A1 sediment decreased from 45.89% of the Daning sediment to 33.14%, while the clay content in the Daning A2 sediment rose to 60.36%.

The morphological differences in each grade sediment are shown in Fig. 5.3 by TEM analysis (Wang et al., 2009). Obviously, a large number of floccules were observed in Daning sediment. After removing carbonates and exchangeable calcium, sediment particles aggregated into a large form, which may be due to the effect of flocculation caused by metal hydroxides and humic acid. The larger size of sediment particles was in accordance with the result of the clay content decreasing. Additionally, there was no apparent difference in floccules. For the Daning A2 sediment, better dispersion occurred and the size of sediment particles obviously decreased, which was also demonstrated by the increase in clay content. Moreover, most floccules disappeared, which made the profile of sediment particles clear. This result indicated that the floccules mainly consisted of organic matter and metal hydroxides. When most organic matter and metal hydroxides were removed, the effect of flocculation was reduced and, consequently, the degree of dispersion was increased.

As shown in Table 5.2, the experimental data of the treated sediment (Daning A1 and Daning A2) was also in good agreement with the power function and Simple Elovich model at the initial P concentration of 1.75 mg/L and 2.54 mg/L. This suggested that the effects of carbonates and calcium, organic matter and metal hydroxides removed from sediments on phosphate adsorption were similar to the original sediment (Daning sediment).

Adsorption isotherms of the treated sediments for phosphate were fitted by the modified Langmuir model and the parameters are shown in Table 5.2. Based on R^2 (>0.97), good fits for Daning A1 and Daning A2 were observed. There was a declining trend in *NAP* and EPC_0. Lower EPC_0 value increased the risk of P release from sediment to water. The maximum adsorption amount (Q_{max}) was 0.205 mg P/g for Daning A1 and 0.083 mg P/g for Daning A2, which was also obviously lower than that for the Daning sediment. After the sediment was chemically extracted by NaOAC+HOAC, the decrease in carbonates and exchange-

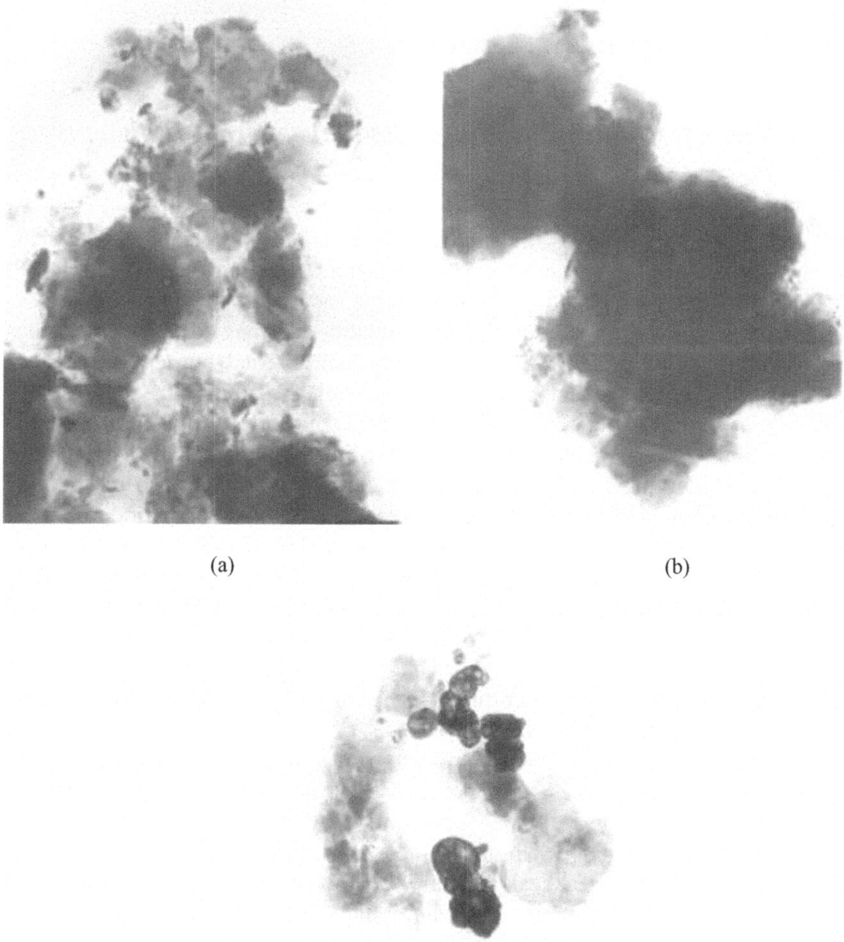

(a) (b) (c)

Fig. 5.3 TEM of each grade sediment extracted from Daning sample (2×10^4 times): (a) Daning sediment sample; (b) Daning A1; (c) Daning A2

able calcium content resulted in the reduction of Ca-bound P. Simultaneously, the larger particles that were formed restrained P adsorption on the sediment. For the Daning A2, an obvious decrease in adsorption capacity for P was attributed to about half of the organic matter and metal hydroxides removed, in spite of the increase in clay content. This result further verified that organic matter and metal hydroxides were the main factors controlling P adsorption on sediments.

5.3 pH

5.3.1 The Effect of pH on Phosphate Release from Sediments

Phosphate may be adsorbed on the sediments which accumulate on the bottom of lakes and reservoirs. These sediments can act as a new pollutant source for the overlying water. One of the most important factors affecting the phosphorus (P) concentration in water was the P release from the sediments. The release processes may have a significant impact on the water quality and may result in continuous eutrophication in eutrophic lakes, especially when external nutrient sources are under control. Therefore, factors affecting the P release from the sediments have been extensively studied and reviewed in past years. Relevant environmental factors mainly include temperature, pH, redox potential, and hydrological conditions. Among them, pH has long been a subject of study. An increase in pH can free P from its binding with ferric complexes due to the competition between hydroxyl ions and the bound P ions (Kim et al., 2003). pH in the overlying water and sediments was a predominant factor as it affected sorption-adsorption, precipitation-solubilization and oxidation-reduction reactions through its control over the concentrations of available iron, aluminum and calcium, thus directly or indirectly changing the aquatic, biological and chemical reactions. pH has become a major factor controlling the availability of the sediment bound P for algal utilization and, consequently, the eutrophication of lakes.

In our study, the effect of phosphate release from Cuntan, Xiaojiang, Daning and Xiangxi sediments from the Three Gorges Reservoir on the Yangtze River was conducted systematically. Release experiments were performed in Pyrex beakers, containing NaCl solution and sediment samples at different pH. 0.5 g of sediments was added to 250 mL Pyrex beakers from 2 to 11. 100 mL of 0.02 mol/L NaCl solution was then added to maintain certain ionic strength. Different amounts of 1 mol/L HCl or 1 mol/L NaOH were added to beakers, respectively, to adjust solution pH. To minimize splashing and evaporation, all the beakers were covered with plastic film. The beakers were shaken on thermostated oscillators for 48 h ((20±1) °C). SRP was then analyzed after the solution was sieved through a 0.45 μm GF/C filter membrane using the ascorbic acid method.

Extreme pH may change the sediment properties, thus affecting the P release from lake sediments. The release rate of P from the four sediments as a function of pH is illustrated in Fig. 5.4. As shown, the rate of P release decreased with the increase in pH in acidic condition. But in alkaline condition, the rate of P release increased with the increase in pH. When pH is neutral, the rate of P release is the lowest. Additionally, more P was released in acidic condition than in alkaline condition.

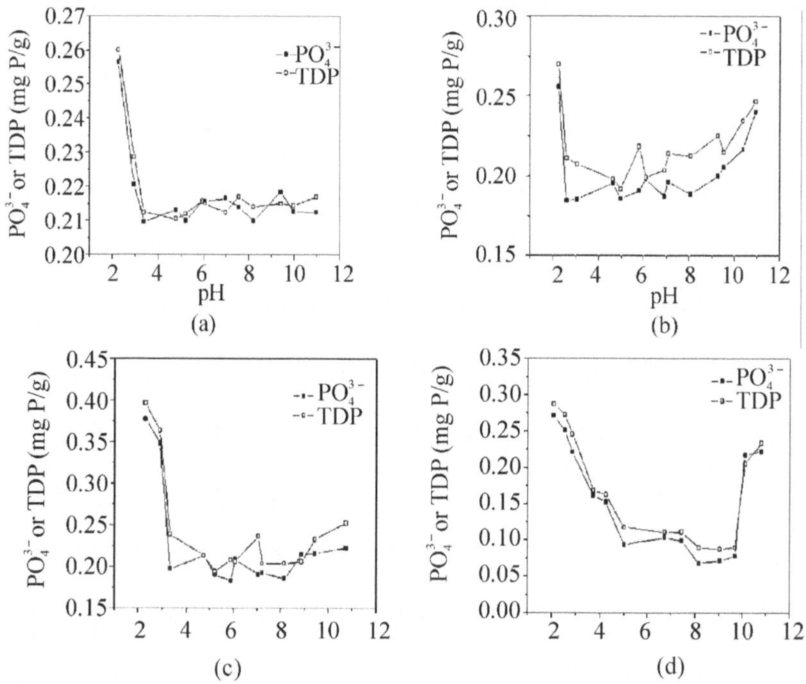

Fig. 5.4 pH dependence of P release on the studied sediment samples. (a) Cuntan; (b) Xiaojiang; (c) Daning; (d) Xiangxi

Those observations suggest that P release from the sediments occurred in both acidic and alkaline conditions, and the acidic condition was more favorable. A neutral condition was the least favorable. The effect of pH on P release was mainly reflected through the P speciation in combination with metals such as Fe, Al, and Ca (Kim et al., 2003). The NaOH-P represented P bound to metal oxides (mainly Al and Fe), and was exchangeable with OH^- and other inorganic P compounds soluble in bases. NaOH-P can be used for the evaluation of algal available P, and can be released for the growth of phytoplankton (Ting and Appan, 1996). HCl-P represented the P fraction sensitive to low pH, and was assumed to consist mainly of apatite P, P bound to carbonates and traces of hydrolysable organic P. Ca-P was a relatively stable fraction, and was attributed to the permanent burial of P in sediments.

For the sediments in our study, inorganic P was the major fraction of total P (accounted for 59.29%–78.82% of total P), wherein the content of calcium bound P (Ca-P) was higher than that of Fe/Al oxides bound P (Fe/Al-P). So higher P release was observed in acid than in alkaline conditions due to the dissolution of Ca-P at lower pH (Fig. 5.5). The lowest release amount of P occurred in neutral conditions. Thus, acidification of the overlying water enhanced the risk of eutrophication in these main branches of the Three Gorges Reservoir.

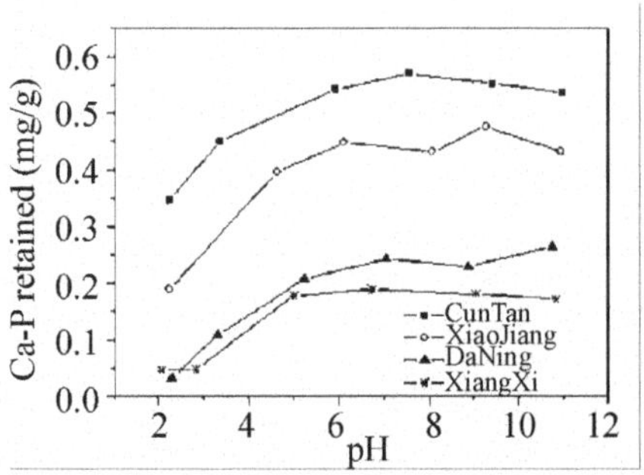

Fig. 5.5 pH dependence of residual Ca-P after P release from the studied sediment samples

5.3.2 The Effect of pH on Phosphate Adsorption on Sediments

As reported, P adsorption on sediments varies with pH in the bulk solution. The effects of pH on P cycling depend in part on the direct, short-term effects of pH on sorption equilibria and kinetics, as well as on the long-term effects of acidification on organic mineralization rates and on sediment composition (Naoml and Patrick, 1991).

In our study, adsorption of P on the sediments from the influx of four main branches (Cuntan, Xiaojiang, Daning and Xiangxi River) of the Three Gorges Reservoir on the Yangtze River varied with the change in the pH values. pH effect on P adsorption (Fig. 5.6) cannot be simply generalized in a wide range of pH observed in our experiments. The figure is mainly separated into three sections by pH 6 and pH 8. The maximum adsorption capacity of various sediments for P occurred in the pH values between 6 and 8. The adsorption capacity followed the order: Xiaojiang> Daning > Cuntan > Xiangxi. As pH<6, the sediment surfaces would compete with a large number of H^+ to trap soluble reactive P (SRP). Whereas as pH>8, OH^- with high free concentration could overcome SRP to be adsorbed onto sediment surfaces. So the adsorption amount of P decreased rapidly when pH<6 or pH>8.

When the curves showing pH effects go below the pH axis, the P preadsorbed will be released and then negative NSP can be observed. The fact that Xiangxi sediment has a greater trend to release adsorbed P when pH is below the pH axis is interpreted by the speculation that adsorption on Xiangxi sediment is more reversible than on other sediments, and meanwhile Xiangxi sediment has more pre-adsorbed P to release.

The general trend that zeta potential decreases with the increase in pH is demonstrated by curve zeta in Fig. 5.6. The result indicates that P sorption may change the surface charge but is not the significant factor. In fact, P adsorption on sediments was shown to be chemical in nature, meaning the main adsorption force comes from chemical affinity (or complexation) rather than from electrostatic interaction on the sediment–water interface. In particular, a strong contribution to the P binding comes from a ligand-exchange process on the Me–OH2 and Me–OH sites. Therefore, it is believed that P sorption, as a function of pH, is more likely to be determined by the stability of phosphate surface complexes than the electrostatic attraction/repulsion from the surface charge (Nemeth et al., 1998).

The effect of pH on P adsorption by Daning sediment and the treated sediment (Daning A1 and Daning A2) was illustrated in Fig. 5.6. The adsorption amount of P by different sediments was in the order: Daning > Daning A1 > Daning A2. That is, the removal of carbonates and calcium, organic matter and metal hydroxides from sediments decreases the capacity of sediments for phosphate adsorption. This result is consistent with that mentioned above.

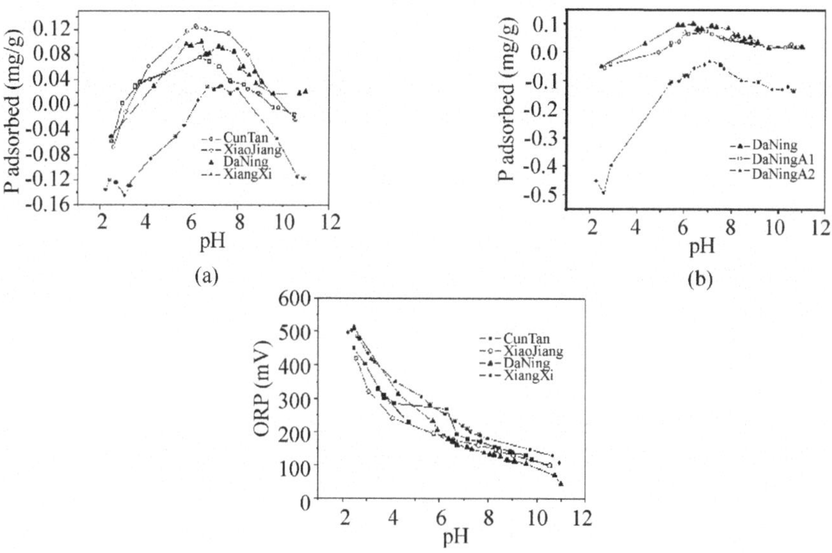

Fig. 5.6 P adsorption on the studied sediment samples at different pH. (a) pH dependence of P adsorption on the studied sediment samples; (b) pH dependence of ORP of the sediment samples

5.4 Temperature

The water temperature of the lakes and reservoirs typically oscillates in diurnal and seasonal cycles in response to air temperature on the sediment–water exchange of phosphate should be considered, which has been studied by many researchers. For example, an increase in adsorption of phosphate with the temperature increasing from 10 to 35 °C has been reported in sediments of Taihu Lake in the lower Yangtze delta region of China (Jin et al., 2005). The effect of temperature on phosphate sorption on marine sediments was studied by Zhang and Huang (2011). They found that more phosphate will be released from suspended particles to ambient water in winter than in summer.

In our study, the adsorption isotherms of phosphate on sediments from Cuntan, Xiaojiang, Daning and Xiangxi sediments from the Three Gorges Reservoir on the Yangtze River at 278, 293 and 303 K were studied. In this section, the adsorption isotherms of phosphate on Cuntan sediment at different temperatures are illustrated in Fig. 5.7. The results show that the adsorption process of phosphate on the sediments is the endothermic reaction as the phosphate adsorption rate increased along with the increasing temperature. Temperature and DO concentration affected phosphate release and adsorption. With an increase in temperature the DO concentration often decreases. All these promote phosphate adsorption. This study only considered the temperature effect.

The adsorption isotherms of the four sediments from Cuntan, Xiaojiang, Daning and Xiangxi sediments from the Three Gorges Reservoir on the Yangtze River at 278, 293 and 303 K were regressed by the Langmuir model and modified Langmuir model, respectively. The veracity of those results was evaluated by R^2 (Table 5.5). The results showed that the adsorption isotherms of phosphate on the four different sediments were more fitted to the modified Langmuir model (R^2>0.95). Theoretically, the maximum adsorption amount ($Q_{\max}$) at 278, 293 and 303 K was 0.284–0.574 mg P/g for Xiaojiang sediment, 0.156–0.453 mg P/g for Daning sediment, 0.09–0.177 mg P/g for Cuntan sediment, and 0.11–0.158 mg P/g for Xiangxi sediment.

Thermodynamic parameters of the adsorption of phosphate on the different trophic sediments, including ΔH^0 heat of adsorption and ΔS^0 entropy change, were calculated using the following equations:

$$\Delta G^0 = \Delta H^0 - T\Delta S^0 \tag{5.9}$$

$$\ln K = \Delta H^0 / RT + \Delta S^0 / R \tag{5.10}$$

where ΔH^0 is the heat of adsorption (kJ/mol), ΔS^0 is the entropy change (kJ/(K·mol)), ΔG^0 is the free energy change (kJ/mol), T is the absolute temperature (K), R is the gas constant (kJ/(K·mol)) and K is a constant related to the adsorption energy (mol^2/kJ^2).

The values of the thermodynamic parameters are shown in Table 5.4. The values of ΔH^0 are obtained from the slope of each plot. The results show that the adsorption of phosphate on the four sediments is an endothermic process. This indicates the spontaneity of the process, the adsorption favored at higher temperatures. The values of ΔH^0 for adsorption of phosphorus by Cuntan, Xiaojiang, Daning and Xiangxi sediments were 18.92, 25.49, 46.63 and 31.88 kJ/mol, respectively. That is, the value of ΔH^0 for adsorption of phosphorus by Daning sediment was higher than that of Cuntan, Xiaojiang and Xiangxi sediments. In this study, the physical meaning of ΔH^0 is the heat of adsorption change of the per mol phosphate distributing from unit water to unit sediment. So the adsorption process of phosphate on the sediments from the Daning sediment was easier than that from the Cuntan, Xiaojiang, and Xiangxi sediments, and was less affected by temperature.

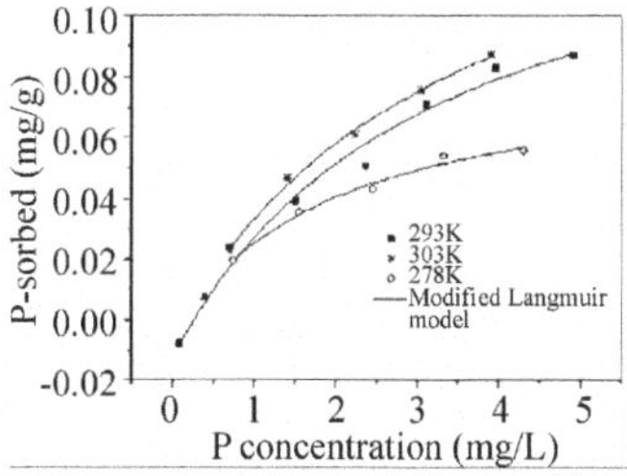

Fig. 5.7 The adsorption isotherm of P on Cuntan sediment at different temperatures

Table 5.4 Langmuir parameters for P adsorption on the studied sediment samples at different temperatures and standard adsorption heat

Samples	T (K)	$Q_{\max}$ (mg/g)		b		R^2		$\ln K$~$1/T$	ΔH^0 (kJ/mol)
		Langmuir model	Modified Langmuir model	Langmuir model	Modified Langmuir model	Langmuir model	Modified Langmuir model		
	278	0.092	0.090	2.640	1.844	0.931	0.998		
Cuntan	293	0.197	0.165	4.935	3.186	0.938	0.986	$\ln K=-2275.4/T+8.832$ $R^2=0.95$	18.92
	303	0.217	0.177	6.930	3.544	0.945	0.997		
	278	0.310	0.284	2.648	1.623	0.921	0.998		
Xiaojiang	293	0.427	0.402	3.974	3.421	0.918	0.972	$\ln K=-3065.8/T+11.562$ $R^2=0.94$	25.49
	303	0.446	0.574	2.203	3.905	0.935	0.997		
	278	0.158	0.156	2.079	1.511	0.918	0.996		
Daning	293	0.220	0.358	3.239	5.743	0.911	0.959	$\ln K=-5608.4/T+20.67$ $R^2=0.95$	46.63
	303	0.237	0.453	1.514	7.573	0.923	0.998		
	278	0.120	0.110	5.789	1.109	0.926	0.989		
Xiangxi	293	0.200	0.150	6.807	1.713	0.901	0.967	$\ln K=-3834.4/T+13.822$ $R^2=0.92$	31.88
	303	0.364	0.158	19.56	3.628	0.942	0.999		

5.5 Ionic Strength

Three different ionic strengths, 0.001, 0.01 and 0.1 mol/L KCl, were used in phosphate adsorption experiments on Daning sediment and treated sediments and the results are compared in Fig. 5.8. All of the adsorption measurements on Daning sediment and treated sediments were modeled empirically using a modified Langmuir-type adsorption isotherm. It was evident that the modified Langmuir model had a good fit with the experimental data for the sediments. Phosphate sorption on the different sediments increased with the decrease in salinity, which may be due to competitive effect. This result was consistent with others (Zhang and Huang, 2011). However, some researchers found the contrary effect (Wang et al., 2006). That is, the phosphate sorption rate increased with an increase in ionic strength. In their experiments, as ionic strength increased from 0.001 to 0.01 mol/L KCl the sorption capacity increased 20% at a constant pH (6–7). This increase can be attributed to the macromolecular phosphate configuration (Wang et al., 2006): In high ionic strength solution, the negative charges on phosphate macromolecules were well screened and consequently the phosphate molecules wound up like random coils. In this more compact configuration, more phosphate can be sorbed onto a given area of sediment surface. As ionic strength decreased, the unscreened negatively charged function groups repelled each other and consequently the molecules spread out as flexible,

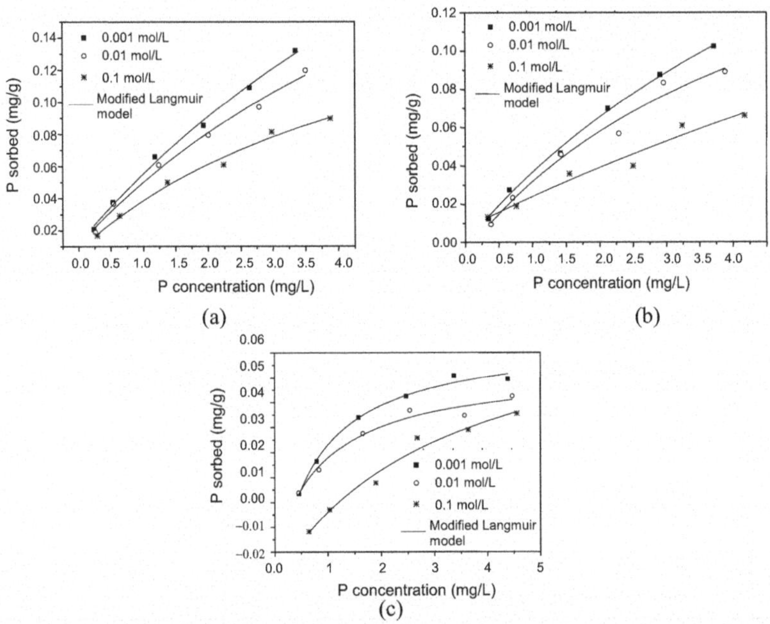

Fig. 5.8 The adsorption isotherm of P on Daning sediment (a) and its treated sediments A1 (b), A2 (c) at different inoic strength

linear polymer molecules. In this configuration the molecules are more spaced out on the surface, resulting in lower sorption density. This suggests that organic matter was one of the most important factors affecting phosphate sorption at the sediment–water interface.

References

Andersen TK, Jensen MH (1983) Nitrogen transformations in coastal sediments, investigated by use of the acetylene inhibition technique with special reference to diurnalvand seasonal variations in denitrification rates. MSc thesis, University of Aarhus, Denmark.

Bai XH, Zhong WG, Chen QY (2002) Effects of sediment and its control in the restoration of polluted urban river. J Environ Sci-China 23(1):89-92.

Battistoni P, Fava G, Ruello ML (1993) Heavy metal shock load in activated sludge uptake and toxic effects. Wat Res 27(5):821-827.

Bergv, Neumann UT, Donnert D, Nüesch R, Stüben D (2004) Sediment capping in eutrophic lakes-efficiency of undisturbed calcite barriers to immobilize phosphorus. Appl Geochem 19:1759-1771.

Boatman CD, Murray JW (1982) Modeling exchangeable NH_4^+ adsorption in marine sediments: Process and controls of adsorption. Limnol Oceanogr 27:99-110.

Borgnino L, Avena M, De Pauli C (2006) Surface properties of sediments from two Argentinean reservoirs and the rate of phosphate release. Wat Res 40:2659-2666.

Brinkman AG (1993) A double-layer model for ion adsorption onto metal oxides, applied to experimental data and to natural sediments of Lake Veluwe, The Netherlands. Hydrobiol 253:31-45.

Bubba DM, Arias CA, Brix H (2003) Phosphorus adsorption maximum of sands for use as media in subsurface flow constructed reed beds as measured by the Langmuir isotherm. Water Res 37:3390-3400.

Burkholder JM, Tomasko DA, Touchette BW (2007) Seagrasses and eutrophication. J Exp Mar Biol Ecol 350(1-2):46-72.

Burt TP, Heathwaite AL, Trudgill ST (1993) Nitrate: Process, Patterns and Management. Wiley, New York.

Carpenter SR, Caraco NF, Correll DL, et al. (1998) Nonpoint pollution of surface waters with phosphorus and nitrogen. Ecol Appl 8:559-568.

Cloern JE (2001) Our evolving conceptual model of the coastal eutrophication problem. Mar Ecol Prog Ser 210:223-253.

Crosby SA, Millward GE, Butler EI (1984) Kinetics of phosphate adsorption by iron oxyhydroxides in aqueous systems. Estuar Coast Shelf Sci 19:257-270.

Fan CX, Morihiro A (1997) Effects of aerobic and anaerobic conditions on exchange of nitrogen and phosphorus across sediment–water interface in lake Kasumigaura. Chinese J Geophys 9(4):337-342.

Fan CX, Zhang, L, Bao XM, et al. (2004) Migration mechanism of biogenic elements and their quantification on the sediment–water interface of Lake Taihu: Biochemical thermodynamic mechanism of phosphorus release and its source-sink transition. J Lake Sci 16(1):10-20.

Fedro U, González T, Herrera-Silveira JA, et al. (2007) Water quality variability and eutrophic trends in karstic tropical coastal lagoons of the Yucatán Peninsula. Estuar Coast Shelf Sci 76(2):418-430.

Focht DD, Verstraete W (1977) Biochemical ecology of nitrification and denitrification. Adv Microbiol Ecol 1:135-214.

Gardolinski PCFC, Worsfold PJ, McKelvie ID (2004) Seawater induced release and transformation of organic and inorganic phosphorus from river sediments. Wat Res 38(3):688-692.

Goreau TJ, Kaplan WA, Wofsy SC, McElroy MB, Valois FA, Watson SW (1980) Production of NO_2 and N_2O by nitrifying bacteria at reduced concentrations of oxygen. Appl Environ Microbiol 40:526-32.

Helder W, De Vries RTP (1983) Estuarine nitrite maxima and nitrifying bacteria (Ems-Dollard Estuary). Neth J Sea Res 17:1-18.

Hou LJ, Liu M, Jiang HY (2003) Ammonia nitrogen adsorption isotherm of tidal flat surface sediments from the Yangtze estuary. Environ Chem 22(6):568-572.

House WA, Denison FH (2000) Factors influencing the measurement of equilibrium phosphate concentrations in river sediments. Water Res 34:1187-1200.

Hu ZQ, Kartik C, Domenico G, et al. (2003) Impact of metal sorption and internalization on nitrification inhibition. Environ Sci Technol 37:728-734.

Hynes RK, Knowles R (1984) Production of nitrous oxide by Nitrosomonas europaea: effects of acetylene, pH and oxygen. Can J Microbiol 30:1397-1404.

Jiang XX, Ruan XH, Xing YN, et al. (2007) Effects of nutrient concentration and do status of heavily polluted urban stream water on nitrogen release from sediment. J Environ Sci-China 28(1):87-91.

Jiao LX (2007) Nitrogen forms characteristic in the sediments from the shallow lakes and functions in biogeochemical cycling. MSc thesis, Inner Mongolia University of Science and Technology, China.

Jiao NZ, Zhang Y, Zeng YH (2007) Ecological anomalies in the East China Sea: Impacts of the Three Gorges Dam? Water Res 41:1287-1293.

Jin X, Wang S, Pang Y, Zhao H, Zhou X (2005) The adsorption of phosphate on different trophic lake sediments. Colloid Surf A 254:241-248.

Jones RD, Hood MA (1980a) Interaction between an ammonium oxidizer, Nitrosomonas sp. and two heterotrophic bacteria, Nocardia atlantica Pseudomonas sp.: a note. Microbiol Ecol 6:271-276.

Jones RD, Hood MA (1980b) Effects of temperature, pH, salinity and inorganic nitrogen on the rate of ammonium oxidation by nitrifiers isolated from wetland environments. Microbiol Ecol 6:339-347.

Jorgensen BB, Revsbech NP (1989) Oxygen uptake, bacterial distribution and carbon-nitrogen-sulfur cycling in sediments from the Baltic Sea-North Sea transition. Ophelia 31:51-72.

Juliastuti SR, Baeyens J, Creemers C (2003) Inhibition of nitrification by heavy metals and organic compounds: the ISO 9509 test. Environ Eng Sci 20:79-90.

Ku WC, Digiano FA, Feng TH (1978) Factors affecting phosphate adsorption equilibria in lake sediments. Wat Res 12:1069-1074.

Li WH, Chen YX, Sun JP (2003) Influence of different dissolved oxygen (DO) Amounts on released pollutants from sediment to overlying water. J Agro-Environ Sci 22(2):170-173.

Lijklema L (1980) Interaction of orthophosphate with iron (III) and aluminum hydroxides. Environ Sci Technol 14:537-540.

Liu PF, Chen ZL, Liu J (2002) Study on effects of salinity and pH on NH_4^+ release in east Chongming tidal flat sediment. J Shenyang Jianzhu University 21(5):271-274.

Lopez P, Lluch XM, Vidal JA, Morgui (1996) Adsorption of phosphorus on sediments of the Balearic Islands (Spain) related to their composition. Estuar Coast Shelf Sci 42:185-196.

Mackin JE, Robert CA, et al. (1984) Ammonium adsorption in marine sediments. Limnol Oceanogr 29(2):250-257.

Makten CS, Rosenfeld JK (1978) Interstitial water chemistry of anoxic Long Island Sound sediments. 2. Nutrient regeneration and phosphate removal. Limnol Oceanogr 23:605-617.

Naoml ED, Patrick LB (1991) Phosphorus sorption by sediments from a soft- water seepage lake. 1. An evaluation of Kinetic and Equilibrium Models. Environ Sci Technol 25(3):395-403.

Nemeth Z, Gancs L, Gemes G, Kolics A (1998) pH dependence of phosphate sorption on aluminum. Corros Sci 40:2023-2027.

Nwoke OC, Vanlauwe B, Diels J, Sanginga N, Osonubi O, Merckx R (2003) Assessment of labile phosphorus fractions and adsorption characteristics in relation to soil properties of West African savanna soils. Agr Ecosyst Environ 100:285-294.

Revsbech NP, Jrgensen BB, Brix O (1981) Primary production of microalgae in sediments measured by oxygen microprofile, $H_{14}CO_3$ fixation, and oxygen exchange methods. Limnol Oceanogr 267:17-30.

Rysgaa S, Thastum P, Dalsgaard T, Christensen PB (1999) Effects of salinity on NH_4^+ adsorption capacity, nitrification, and denitrification in Danish estuary sediments. Estuaries 22:21-30.

Seitzinger SP (1991) The effect of ph on the release of phosphorus from Potomac Estuary sediments: implications for the blue-green algal blooms. Estuar Coast Shelf Sci 33(4):409-418.

Sharma B, Ahlert RC (1977) Nitrification and Nitrogen Removal. Wat Res 11:897

Suzuki I, Dular V, Juok SC (1974) Ammonia or ammonium as substrate for oxidation by Nitrosomonas europaea cells and extracts. J Bacteriol 1230:556-558.

Tang HX, Xue HB, Tian BZ (1982) Study on multi-component adsorption model of aquatic sediments with a sequential chemical separation procedure. Acta Scientiae Circumstantiae 2:279-292.

Thomann RV, Mueller JA (1987) Principles of surface water quality modeling and control. Harper and Row, Inc., New York.

Ting DS, Appan A (1996) General Characteristics and Fractions of phosphorus in aquatic sediments of two tropical reservoirs. Water Sci Technol 34:53-59.

Tu C, Zheng CR, Chen KM (2002) Effect of Heavy Metals on Phosphorus Retention by Typic Udic Ferrisols: Equilibrium and Kinetics. Pedosphere 12:15-24.

von Sperling E, da Silva Ferreira AC, Gomes LNL (2008) Comparative eutrophication development in two Brazilian water supply reservoirs with respect to nutrient concentrations and bacteria growth. Desalination 226(1-3):169-174.

Wang HM, Ji YQ, Zhong FY, et al. (2004) Mathematical model of the transport and conversion of triple-nitrogen in a mixed reservoir (lake) under ideal conditions and its solution. J Eng Geol Hydroge 2:35-38.

Wang SR, Jin XC, Bu QY (2006) Effects of particle size, organic matter and ionic strength on the phosphate sorption in different trophic lake sediments. J Hazard Mater 128:95-105.

Wang SR, Jin XC, Zhao HC, Zhou XN, Wu FC (2007) Effect of organic matter on the sorption of dissolved organic and inorganic phosphorus in lake sediments. Colloids Surf A 297:154-162.

Wang Y, Shen ZY, Niu JF, Liu RM (2009) Adsorption of phosphorus on sediments from the Three-Gorges Reservoir (China) and the relation with sediment compositions. J Hazard Mater 162(1):92-98.

Welch EB (1992) Ecological effects of wastewater. Applied Limnology and Pollutant Effects, 2nd ed. Chapman and Hall, London.

Weng YG, Xing Y, Zhang CX (2006) Effects of three heavy metal ions on nitrosification and nitrification in sea water. J Dalian Fisheries University 21(1):51-54.

Wetzel RG (1983) Limnology (2nd ed.). Saunders Philadelphia PA.

Wu JG, Huang JH, Han XG, Xie ZQ, Gao XM (2003) Three-Gorges Dam: experiment in habitat fragmentation? Science 300:1239-1240.

Wu QH, Zeng XY, Huang Y (2005) Influence of DO and Organic Matter on Nitrogen (NH_4^+-N, NO_2-N and NO_3-N)) Releasing in the Sediment of River. Research of Environmental Sciences 18(5):34-39.

Yang W, Wang Y, Guo Y (2008) Test about effect of iron and manganese on nitrogen transformation in underground water. J Shenyang Jianzhu University 4(2):286-290

Zeng H, Song LR, Yu ZG, Chen HT (2006) Distribution of phytoplankton in the Three-Gorges Reservoir during rainy and dry seasons. Sci Total Environ 367:999-1009.

Zhang JZ, Huang XL (2011) Effect of Temperature and Salinity on Phosphate Sorption on Marine Sediments. Environ Sci Technol 45:6831-6837.

Zhao L, Yang XF (2009) Research progress of nitrogen transformation in water eco-environment. Environ Sci-JiangSu S2:025.

Zhou AM, Tang HX, Wang DS (2005) Phosphorus adsorption on natural sediments: Modeling and effects of pH and sediment composition. Water Res 39:1245-1254.

6

Mathematical Modeling and Numerical Simulation

6.1 Overview

Eutrophication has become a major environmental water issue in many parts of the world. Emission of excessive nutrient loads generated from point and non-point pollution sources into water bodies such as lakes, reservoirs and rivers leads to high phytoplankton biomass growth. An overabundance of algal biomass causes water quality problems, like variation in the diurnal oxygen level, oxygen depletion in deep water, unpleasant tastes and odors in the water supply, clogging of filters in water treatment plants, and adverse effects on water-contact sports and recreation. With the rapid development of agriculture and industry, eutrophication has recently been observed in a variety of water bodies in China, including Chaohu Lake (Yin and Hang, 2003), Dianchi Lake (Sun and Guo, 2002), Taihu Lake (Liu and Chen, 2006) and the Miyun Reservoir (Li et al., 2007).

To carry out quantitative prediction of water quality changes and scenario analysis, mathematical models are often used as an effective technical tool. Generally, the mathematical models used in simulating eutrophication issues can be divided into 4 kinds: simple regression model, simple nutrients balance model, water ecological system model and structurally dynamic eutrophication model.

The simple regression model is based on a great amount of statistical analysis of water quality and biological data, most of this used to describe the relationship between algae chlorophyll a and phosphorus or transparency (Vollenweider, 1975; Dillon, 1975). In the 1970s, the Organisation for Economic Co-operation and Development (OECD) organized co-operation research for lake eutrophication in the world. About 200 main lakes were considered, and an obvious relationship between limited productivity of nutrient elements and lake eutrophication status indicators was discovered (Liu et al., 1987). A simple regression model only needs less parameters. It can be used to forecast the rough trend of water quality in a lake or reservoir. But the regression model is derived from a specific research area, so portability is not possible.

The phosphorus balance model is the typical representative of simple nutrients balance models. Many researches have showed that phosphorus is the limiting factor in lake eutrophication. In 1969, Vollenweider developed the total phosphorus material balance model for the first time. Take the lake as a black box, according to the principle of material balance. He assumed that the phosphorus concentration changing over time is equal to the input of phosphorus minus sedimentary phosphorus within a lake and output phosphorus per unit volume (Peng et al., 2007). Later, many similarly improved nutrients balance models were developed based on the phosphorus balance model (Imboden, 1974; Imboden and Gachter, 1978). But the nutrients balance model may only suit small lakes or reservoirs.

The water ecological system model is based on the mass, momentum and energy balance equation, with the dynamic process of various ecological variables as the core, to simulate the variation processes of ecological variables. At present, water ecological system models are the main part of the eutrophication models. Many scholars developed all kinds of complex water ecological system models to simulate the physical, chemical, biological, ecological and hydrodynamic processes of ecological systems (Malmaeus and Håkanson, 2004; George and Michael, 2005; Jørgensen et al., 2002; Koelmans et al., 2001). Among them, the Jørgensen model, CE-QUAL-W2 (Cole and Buchak, 1995), EFDC (the Environmental Fluid Dynamics Code) model (Hamrick, 1992; 1996), and the WASP (Water Quality Analysis Simulation Program) model (Wool et al., 2001) are very popular.

The structurally dynamic eutrophication model was developed originally in the 1980s. It considered the plasticity and variability of a lake's ecosystem. A set of a continuous variation of the parameters and the objective function were used to reflect the adaptation ability of biological ingredients to the changes in the external environment. Exergy was the main objective function of the structurally dynamic eutrophication model, and the model has been successfully applied in many lakes.

Model tools like QUAL2K (Chapra et al., 2007), WASP, CE-QUAL-ICM (Cerco and Cole, 1994), CE-QUAL-W2, MIKE (DHI, 1995; 1999), SMS (the Surface Water Modeling System) (Henderson et al., 2004), and EFDC are extensively used in modeling rivers, lakes, estuaries all over the world. However, most of these models are applicable for small or meso-scale water environments. Numerical modeling for macro-scale water environments is difficult due to complex fluvial geomorphology, meteorological conditions, and fluctuations of point and non-point pollutant emission. In addition, the demand for a vast quantity of data for model calibration and verification also restricts the development and application of water quality models to large scale water bodies.

As one of the biggest water conservancy projects in the world, the construction of the Three Gorges Dam (TGD) has attracted a lot of attention from environmentalists, since the project may cause significant ecological and environmental changes over a wide area (Edmonds, 1992; Wu et al., 2004). The Three Gorges Reservoir Region (TGRR) is affected directly by the reservoir water impoundment and has been researched a lot. A direct consequence of the reservoir impoundment is the decrease in the flow velocity, which causes more sediment settling, slower pollutant transportation and lower DO concentrations in deep water. In some tributaries, the flow velocity decreases to less than 0.01 cm/s in the backwater zones where water containing high total nitrogen (TN) and total phosphorus (TP) enters because of the rising water level (Li et al., 2002). Eutrophication appears frequently in TGRR (MEP, 2004; 2005; 2006; 2007).

The TGRR ranging from Chongqing City to Yichang City, Hubei Province (106°16′ ~111°28′E, 28°56′~31°44′N) (Fig. 6.1) refers to the area affected by the water impoundment of the TGP and covers an area of 5,4000 km^2 (Zhang et al., 2009). Most of this area is mountainous and the west, with an elevation of between 1000–2500 m, is generally higher than the east (about 500–900 m). The tropical monsoon climate of northern Asia characterizes the study area. The annual mean temperature, precipitation, evaporation rate and wind speed are 18°C, 1170 mm, 1300 mm, 1.4 m/s, respectively. In the TGRR, the water resource is quite plentiful with an average flow of 40.56 billion m^3 per year, in which underground runoff accounts for about 21%. The temporal and spatial distribution is very uneven and the northwest is short water resources in spring and winter time. There are more than 44 tributaries whose drainage area is larger than 100 km^2, including Jialing River, Wujiang River and Daning River etc. (Deng, 2007).

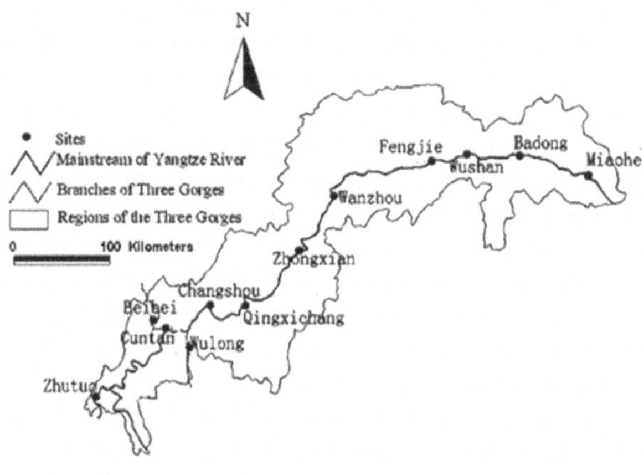

Fig. 6.1 Location of monitoring stations in the TGRR

Modeling research has been undertaken for more than 20 years in the TGRR. In 1991, a report on "The Impact of the Three Gorges Project on the Ecology and Environment" was completed by the Changjiang Water Resources Protection Institute coordinating with the Chinese Academy of Sciences and other institutes (CWRPI and CAS, 1991). In this report, due to the limitation of the technical conditions, qualitative evaluation of the effects of water impoundment on water quality was preliminarily carried out, and only a few empirical formulae and simple mathematical models were used (Yu, 2008). During 1989~2001, using 2-D and quasi 2-D vertically mixed models, Chongqing Environmental Science Institute cooperating with research institutes from the United Kingdom and Denmark launched a series of hydrodynamic and water quality simulations in the Chongqing section. COD, NH_4-N, BOD_5, and other water quality parameters were simulated; however, water environment changes in the entire reservoir area were not involved in their research. In 1996, a systemic research program, Water Pollution Control in the Three Gorges Reservoir Area Program, was launched by the Three Gorges Project Construction Commission of the State Council and several research organizations (Huang et al., 2006). Three topics were included in their program: (1) identification of current pollution loads and forecast of the trend; (2) analysis of water quality change caused by water impoundment; and (3) assigning water environmental capacity of pollutants. For the purpose of predicting the water environment in different impoundment conditions, a 1-D hydrodynamic and water quality model was developed by China Institute of Water Resources and Hydropower Research for the entire reservoir area, from the dam to about 660 kms upstream (Li et al., 2002). Due to restrictions of basic data, the model does not take into account the impact of sediment, and some biochemical processes had to be simplified. For example, phosphorus was treated as a conservative substance. To investigate adsorption and desorption of TP on the sediment and TP, Yu (2008) introduced a water quality model for sediment-laden flow. However, the process was very simple.

In this chapter, a macro-scale one-dimensional integrated model including hydrodynamics, sediment and water quality sub-models was developed, based on former models (EFDC). In the spatial scale, the model included the whole backwater zone of about 660 km, and data obtained in 11 monitoring stations were used for model establishment, calibration and verification. On the temporal scale, the calibration and verification process lasted 2 complete hydrological periods, 2004 and 2006 respectively. In addition, more water variables were simulated and more complicated biochemical processes were considered, for example, diagenesis in the sediment bed, which is an internal source of inorganic phosphorus and nitrogen in the water column. With this model, hydrodynamic and water quality conditions for each section under various scenarios can be obtained (Zhao et al., 2011). The simulation results can be used as boundary conditions when simulating eutrophication in tributaries. Then a 3-D eutrophication model for Daning River confluence mouth using EFDC software was constructed. The calibration period was from February 10, 2006 to November 11, 2006, and the validation period was from January 24, 2007 to October 1, 2007 according to the data availability. The model can simulate the spatial and temporal variation characteristics of water temperature, sediment, DO, TP, TN and Chl-a.

6.2 Mathematical Models and Numerical Simulation

6.2.1 Model Description

The EFDC model, originally developed by John Hamrick at the Virginia Institute of Marine Science (VIMS), is a general purpose 3-D modeling package for flow, pollutant transport and biogeochemical processes in surface water systems including rivers, lakes, estuaries, reservoirs, wetlands and coastal regions (Hamrick, 1992). In addition to hydrodynamic, salinity and temperature transport simulation capabilities, EFDC is capable of simulating cohesive and non-cohesive sediment transport, near field and far field discharge dilution from multiple sources, eutrophication processes, the transport and fate of toxic contaminants in the water and sediment phases, and the transport and fate of various life stages of finfish and shellfish. Special enhancements to the hydrodynamic portion of the model, including vegetation resistance, drying and wetting, hydraulic structure representation, wave-current boundary layer interaction, and wave-induced currents, allow refined modeling of wetland marsh systems, controlled flow systems, and near shore wave induced currents and sediment transport (Park et al., 2000).

The structure of the EFDC model includes four major modules: (1) a hydrodynamic sub-model, (2) a water quality sub-model, (3) a sediment transport sub-model, and (4) a toxics sub-model. The hydrodynamic sub-model, EFDC, was developed by Hamrick (1992; 1996). The model solves the Navier–Stokes equations for a water body with a free surface. The two turbulence parameter transport equations implement the Mellor and Yamada's level 2.5 turbulence closure scheme (Mellor and Yamada, 1982) as modified by Galperin et al. (1988). Both the turbulent kinetic energy and the turbulent length scale are solved using dynamically coupled transport equations. In the vertical direction, sigma coordinates, with the hydrostatic assumption, are used in the model. Horizontally, curvilinear orthogonal grids are used. A multiple-class sediment transport model (Kim et al., 1998; Lin and Kuo, 2003) and a wetting and drying scheme (Ji, 2001) are included in the model. The water quality sub-model (Park and Kuo, 1996; Park et al., 1998) consists of a water column water quality model and a sediment diagenesis model, both linked internally. The water column water quality model simulates the spatial and temporal distributions of 21 state variables in the water column. The simulated kinetic processes in the water quality sub-model (Fig. 6.2) include algal growth, metabolization, predation, hydrolysis, mineralization, nitrification and denitrification. The kinetic formulations in the model are mostly from CEQUAL-ICM (Cerco and Cole, 1993; 1994), with differences listed in (Park and Kuo, 1996; Park et al., 1998; 2000).

A sediment process model (Fig. 6.3), developed by Di Toro and Fitzpatrick, is coupled with the water column water quality model to simulate processes in the sediments and the sediment–water interfaces (Di Toro and Fitzpatrick, 1993).

The EFDC model's hydrodynamic component is based on the 3-D hydrostatic equations formulated in curvilinear-orthogonal horizontal coordinates and a sigma vertical coordinate. According to anelastic approximation and Boussinesq approximation, the governing equations are:

(1) The momentum equations

$$
\begin{aligned}
&\partial_t(mHu)+\partial_x(m_y Huu)+\partial_y(m_x Hvu)+\partial_z(mwu)-(mf+v\partial_x m_y-u\partial_y m_x)Hv \\
&\quad =-m_y H\partial_x(g\zeta+p)-m_y(\partial_x h-z\partial_x H)\partial_z p+\partial_z(mH^{-1}A_v\partial_z u)+Q_u
\end{aligned}
\tag{6.1}
$$

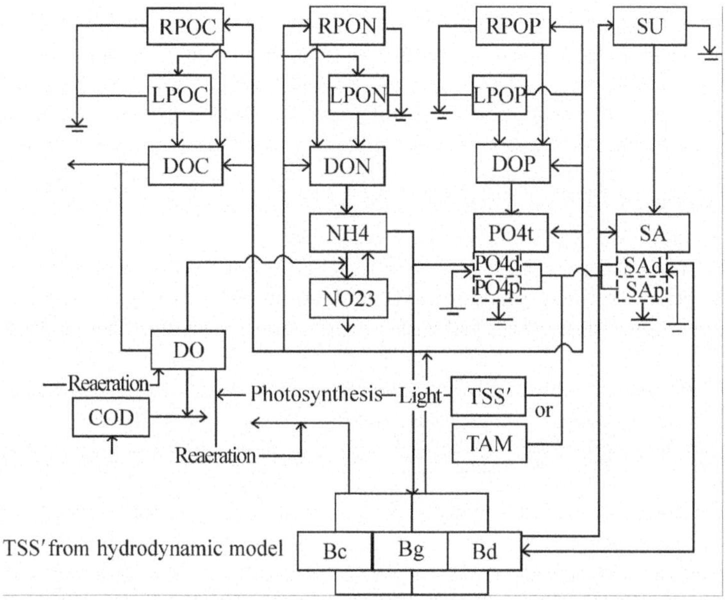

Fig. 6.2 Schematic diagram for the EFDC water column water quality model. RPOC, RPON, RPOP are refractory particulate organic carbon, nitrogen, and phosphorus; LPOC, LPON, LPOP are liable particulate organic carbon, nitrogen, and phosphorus; DOC, DON, and DOP are dissolved organic carbon, nitrogen, and phosphorus; SU and SA are particulate biogenic silica and dissolved available silica; Bc, Bg, and Bd are cyanobacteria, green, diatom algaes; TAM is total active metal; FCB is fecal coliform bacteria

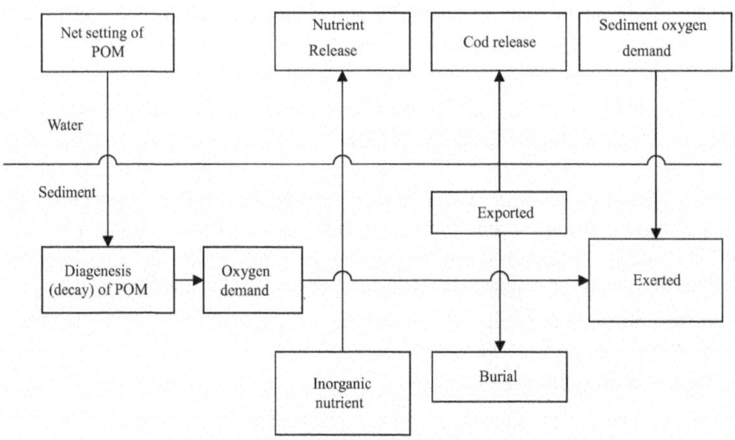

Fig. 6.3 Schematic diagram for sediment process model

$$\partial_t(mHv)+\partial_x(m_yHuv)+\partial_y(m_xHvv)+\partial_z(mwv)+(mf+v\partial_x m_y-u\partial_y m_x)Hu$$
$$=-m_xH\partial_y(g\zeta+p)-m_x(\partial_y h-z\partial_y H)\partial_z p+\partial_z(mH^{-1}A_v\partial_z v)+Q_v \tag{6.2}$$

$$\partial_z p=-gH(\rho-\rho_0)\rho_0^{-1}=-gHb \tag{6.3}$$

(2) The continuity equation

$$\partial_t(m\zeta)+\partial_x(m_yHu)+\partial_y(m_xHv)+\partial_z(mw)=0 \tag{6.4}$$

$$\partial_t(m\zeta)+\partial_x(m_yH\int_0^1 u\mathrm{d}z)+\partial_y(m_xH\int_0^1 v\mathrm{d}z)=0 \tag{6.5}$$

(3) The state equation

$$\rho=\rho(p,S_a,T) \tag{6.6}$$

(4) The transport equation for salt and temperature

$$\partial_t(mHS_a)+\partial_x(m_yHuS_a)+\partial_y(m_xHvS_a)+\partial_z(mwS_a)=\partial_z(mH^{-1}K_v\partial_z S_a)+Q_{aS} \tag{6.7}$$

$$\partial_t(mHT)+\partial_x(m_yHuT)+\partial_y(m_xHvT)+\partial_z(mwT)=\partial_z(mH^{-1}K_v\partial_z T)+Q_T \tag{6.8}$$

(5) The transport equation for suspended sediments

$$\partial_t\left(mHS\right)+\partial_x\left(m_yHuS\right)+\partial_y\left(m_xHvS\right)+\partial_z\left(mwS\right)-\partial_z\left(mw_{sj}S\right)$$
$$=\partial_x\left(\frac{m_y}{m_x}HK_H\partial_x S\right)+\partial_y\left(\frac{m_x}{m_y}HK_H\partial_y S\right)+\partial_z\left(m\frac{K_v}{H}\partial_z S\right)+Q_S \tag{6.9}$$

(6) The water quality governing equation

$$\partial_t(m_xm_y\mathrm{H}C)+\partial_x(m_yHuC)+\partial_y(m_xHvC)+\partial_z(m_xm_ywC)$$
$$=\partial_x(\frac{m_yHA_x}{m_x}\partial_x C)+\partial_y(\frac{m_xHA_y}{m_y}\partial_y C)+\partial_z(m_xm_y\frac{A_z}{H}\partial_z C)+m_xm_yHS_c \tag{6.10}$$

where, u, v and w are the velocity components in the curvilinear, orthogonal coordinates and sigma vertical coordinate x, y and z; m_x and m_y are the square roots of the diagonal components of the metric tensor; $m = m_x m_y$ is the Jacobian of the metric tensor determinant; A_v is the vertical turbulent viscosity; f is the Coriolis parameter; H is depth of water body; ζ is elevation of water surface; p is the physical pressure; ρ is the density; S_a is the salinity; T is the temperature; Q_u and Q_v are momentum source-sink terms; Q_{sa} represents external sources and sinks for salt; Q_t represents external sources and sinks for temperature; S is the suspended sediment concentration; K_v and K_H are vertical and horizontal turbulent diffusion coefficients; w_{sj} is a positive settling velocity; Q_s represents external sources and sinks for sediment; C is the concentration; A_x, A_y and A_z are turbulent diffusion coefficients in x, y and z; S_c represents external sources and sinks for solution. In this study, it is assumed that C equals 0 and the density and temperature are constants. The system of Eqs. (6.1)–(6.10) provides a close system for the variables u, v, w, p, ρ, S_a, T, ζ, S and C.

The numerical scheme employed in EFDC to solve the equations of motion uses second order accurate spatial finite difference. The model's time integration employs a second order accurate three time level, finite difference scheme with an internal-external mode splitting procedure. The EFDC model applies drying and wetting in shallow areas by a mass conservative scheme, by which the astringency of the model is improved. The EFDC theory in detail can be found in Hamrick (1992).

6.2.2 Model Results Evaluation

The simulation results were evaluated using pre-defined statistics parameters, as follows:

(1) Average Error (AE)

$$AE = \frac{\sum_{i=1}^{N}(O_i - X_i)}{N} \tag{6.11}$$

(2) Relative Error (RE)

$$RE = \frac{\sum_{i=1}^{N}\left|O_i - X_i\right|}{\sum_{i=1}^{N}O_i} * 100 \tag{6.12}$$

(3) Average Absolute Error (AEE)

$$AAE = \frac{\sum_{i=1}^{N}\left|O_i - X_i\right|}{N} \tag{6.13}$$

(4) Root Mean Square (RMS) Error

$$RMS = \sqrt{\frac{\sum_{i=1}^{N}(O_i - X_i)^2}{N}} \tag{6.14}$$

where O is observed value (measured data in this manuscript); X is corresponding model value in space and time; and N is number of valid data/model pairs.

6.3 A Macro-Scale One-Dimensional Integrated Model for the Three Gorges Reservoir Area

6.3.1 Data Acquisition and Preprocessing

The basic data sets required to develop the models included topography, hydrology, meteorological and water quality data. A topographic map, with a resolution of 1:25000, was used to delineate the channel character, and bathymetry data along the river was from the literature (Huang et al., 2006). The water surface elevation and flow rate data from 2003–2007 of 9 monitoring stations in the main stream of the Yangtze River including Zhutuo, Cuntan, Changshou, Qingxichang, Zhongxian, Wanzhou, Fengjie, Wushan, and Badong Stations, one station, Beibei Station, located in the Jianglingjiang River and one station, Wulong, located in the Wujiang River, were collected from the Bureau of Hydrology, Changjiang Water Resources Commission. Meteorological data of the reservoir region including atmospheric pressure, air temperature, precipitation, evaporation, relative humidity, wind speed and solar radiation in 2003–2007 were downloaded from the website of the Chingqing Meteorological Bureau (http://www.climate.cq.cn/). Water quality data of COD, DO, NH_4-N, NO_3-N, TP and TN in 2003–2007 in the 11 stations were obtained from the Bureau of Hydrology, Changjiang Water Resources Commission. Daily sediment data in this period of 2003–2007 in the monitoring stations of Zhutuo, Cuntan, Qingxichang and Wanzhou were obtained from the same agency.

Nutrient loads of industrial and municipal point pollution sources in each county and district in 2002 in the TGRR were obtained from China Environmental Science Research Institute (CRAES), and data in 2004–2007 were collected from the Three Gorges Bulletins (MEP, 2004; 2005; 2006; 2007). Non-point source nutrient emission data were obtained from the statistics of Cao (2006) and Zhen et al. (2009).

6.3.2 Model Configuration

Boundary fixed orthogonal curvilinear grids are used for delineating modeled regions in the EFDC model. SEAGRID software of the U.S. Geological Survey (Denham, 2006) was applied as the third utility to create 720 active cells for the whole reservoir area. GEFDC software was used to create dxdy.inp and lxly.inp files (Hamrick, 1992). For the initial condition, water surface elevation, cohesive sediment and other water quality variables for each active cell were interpolated according to monitoring data along the Yangtze River. Due to cohesive sediment with particle sizes less than 100μm accounting for about 90% of the total sediment particles, non-cohesive and bedload were not activated in this model. In view of collected data, only DO, NH_4-N, NO_3-N, organic nitrogen (ON), orthophosphate, organic phosphorus (OP), and COD were modeled in the water quality sub-model. Condition variables not obtained from regular samples were derived according to analysis and test results of Wang et al. (2008). As to the boundary condition, Zhutuo Stations, Beibei Station and Wulong Station were treated as the inflow boundary condition and the open boundary was the downstream Three Gorges Dam. Meteorological conditions were used as the driving function for the model. Default parameters were used initially for model running, and were adjusted during calibration with data in 2004. Model validation was conducted with data from 2006.

6.3.3 Parameters Estimation

The bottom roughness coefficient is adjusted frequently in hydrodynamic modeling, which ranges from 0.01 m to 0.1 m (Ji, 2008). In this study, the bottom roughness coefficient was 0.027 m, determined through model results comparison. In the sediment sub-model of the EFDC version maintained by Dynamic Solution, LLC (Craig, 2009), the setting velocity, Tau critical-deposition below which deposition occurs, Tau critical-erosion above which erosion occurs, and reference surface erosion rate are the main parameters needed to be adjusted for non-cohesive sediment. Tau critical-deposition and Tau critical-erosion, with a unit of m^2/s^2, are computed by normalizing the shear stress by the water density. In this model, the setting velocity, Tau critical-deposition, Tau critical-erosion and reference surface erosion rate were 0.00002 m/s, 0.003 m^2/s^2, 0.005 m^2/s^2, and 0.0005 m^2/s^2, respectively, for the cohesive sediment. In the water quality sub-model, DO concentration is affected by the COD and NH_4-N decay rates, so a maximum nitrification rate of 0.1 g N/(m^3·day) and a COD decay rate of 0.025 day^{-1} were adopted. A value of 0.1 was assigned for the partition coefficient for sorbed/dissolved PO_4 to TSS. Used parameters are listed in Table 6.1.

Table 6.1 Parameters used in the Three Gorges Model

Parameters	Unit	Value
Roughness	-	0.027
IC bed mass	g/m^2	10000
Specific volume	m^3/g	4.4E-07
Specific gravity	-	2.25
Settling velocity	m/s	0.00002
Tau critical-deposition	m^2/s^2	0.003

Parameters	Unit	Value
Tau critical-erosion	m^2/s^2	0.005
Ref. surf erosion rate	$g/(m^2 \cdot s)$	0.0005
Maximum nitrification rate	$g\ N/(m^3 \cdot day)$	0.1
Oxygen half-sat constant for nitrification	mg/L	1
NH_4 half-sat constant for nitrification	$g\ N/m^3$	1
Reference temperature for nitrification	°C	20
Suboptimal temperature coefficient for nitrification	-	0.069
Superoptimal temperature coefficient for nitrification	-	0.0045
Minimum hydrolysis rate of RPON	$1\ day^{-1}$	0.005
Minimum hydrolysis rate of LPON	$1\ day^{-1}$	0
Minimum mineralization rate of DON	$1\ day^{-1}$	0.05
Partition coefficient for sorbed/dissolved PO_4 (to TSS or TAM)	-	5
Minimum hydrolysis rate of RPOP	$1\ day^{-1}$	0.005
Minimum hydrolysis rate of LPOP	$1\ day^{-1}$	0
Minimum mineralization rate of DOP	$1\ day^{-1}$	0.05
Reaeration constant	-	3.933
Temperature rate constant for reaeration	-	1.024
Reaeration adjustment factor	-	5
Oxygen half-sat constant for COD decay	mg/L O_2	1
Settling velocity for RPOM	m/day	0.01
Reference temperature for COD decay	°C	20
Temperature rate constant for COD decay	-	0.041

6.3.4 Model Calibration

Based on the availability of the measured data, the data obtained in Cuntan, Qingxichang and Wanzhou Monitoring Stations (Fig. 6.1) were selected for parameter calibration. The period was 366 days, from January 1 to December 31, 2004. The time step was 60 s, and it took about 1 h to complete model implementation. The simulation results were compared with measured data in the following ways: the time series comparison in each station and the longitudinal profile comparison in the whole reservoir region.

- *Time Series Comparison*

For the purpose of simplicity, this chapter only gives the time series comparison in the Wanzhou Station. The compared constituents included: water surface elevation, flow rate, sediment, T, DO, NH_4-N, NO_3-N, TN, COD, and TP. Fig. 6.4 compares the simulation results and measured data. Table 6.2 lists statistical results on the above-mentioned 9 variables except for flow rate in three stations, because of the close relationship between flow rate and water surface elevation.

The simulated and measured water surface elevation matches very well before and after the high flow period, and the simulation results are a little greater than the corresponding measured values during the high flow period (Fig. 6.4a). For the flow rate (Fig. 6.4b), there is a great change during a hydrological period from several thousand m^3/s to more than 60,000 m^3/s. The model simulated the flow rate satisfactorily in the low flow period, after which there are some higher measured values. This may be caused by abundant inflows from other ignored tributaries in the reservoir region such as the Yulin River, the Daning River, and the Xiangxi River, etc, and the extreme peak value of the measured data cannot be reproduced in the model process, but generally the errors are very small, as shown in Table 6.2.

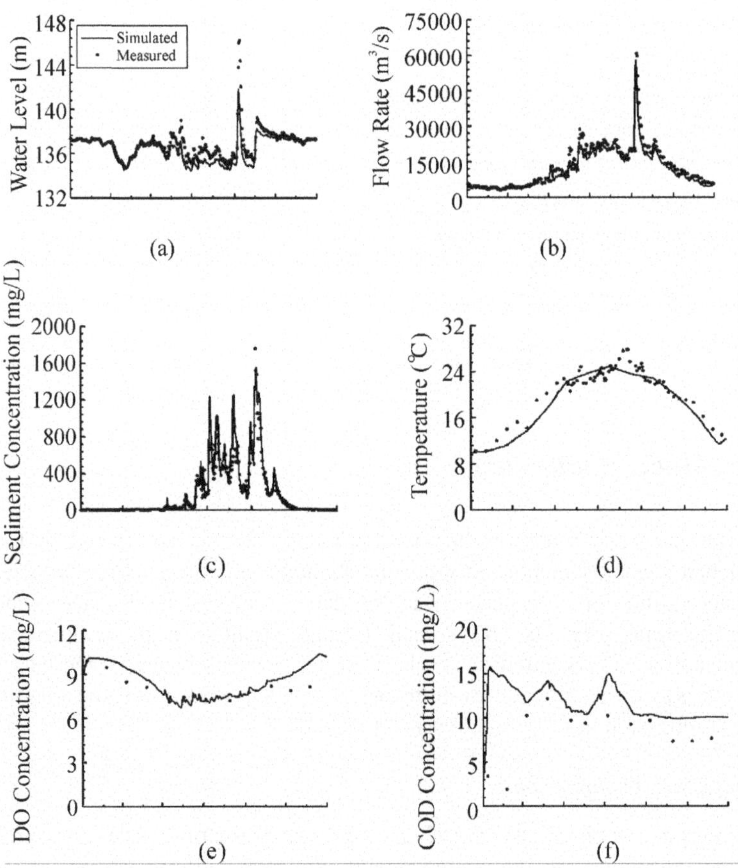

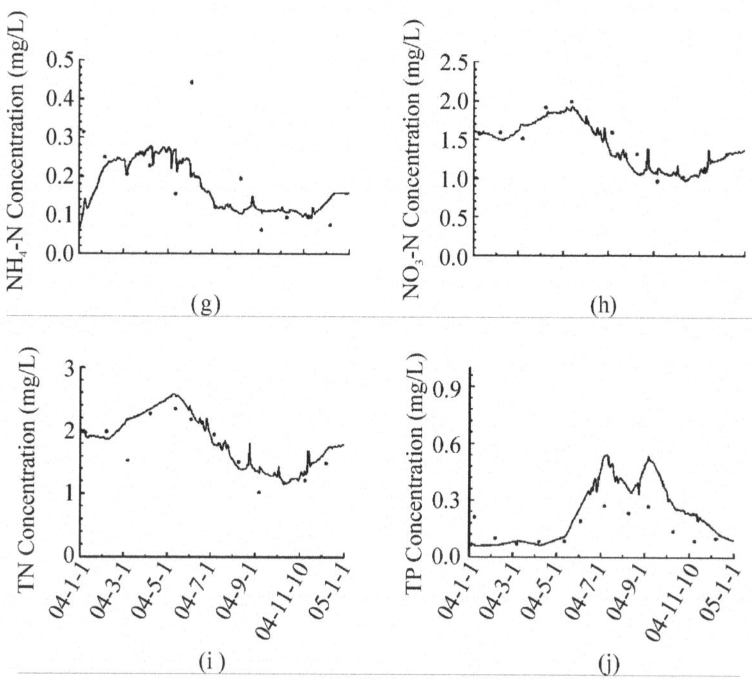

Fig. 6.4 Time series comparsion of measured and simulated hydrological, sediment and water quality results in the Wanzhou Station, 2004

Table 6.2 Simulated results analysis in Cuntan, Qingxichang and Wanzhou Stations, 2004

Station ID	Parameter	Pairs	Data Avg.	Model Avg.	AE	RE	AAE	RMS
Cuntan	Water Surface (m)	366	162.40	161.63	−0.77	0.49	0.79	1.02
Qingxichang	Water Surface (m)	366	141.62	142.37	0.76	0.56	0.79	0.94
Wanzhou	Water Surface (m)	366	136.83	136.44	−0.40	0.29	0.40	0.65
Cuntan	Sediment (mg/L)	366	299.46	257.03	−42.44	17.59	52.67	89.33
Qingxichang	Sediment (mg/L)	366	244.19	216.85	−27.34	21.10	51.51	105.57
Wanzhou	Sediment (mg/L)	366	166.21	167.82	1.61	24.62	40.91	85.45
Cuntan	Temperature (°C)	55	21.46	21.19	−0.27	4.88	1.05	1.41
Qingxichang	Temperature (°C)	49	21.41	20.57	−0.83	8.19	1.75	2.14
Wanzhou	Temperature (°C)	66	21.71	21.30	−0.40	5.57	1.21	1.58
Cuntan	Dissolved Oxygen (mg/L)	12	9.01	8.79	−0.22	5.97	0.54	0.64
Qingxichang	Dissolved Oxygen (mg/L)	12	8.98	8.82	−0.16	7.01	0.63	0.77
Wanzhou	Dissolved Oxygen (mg/L)	12	8.07	7.90	−0.17	5.84	0.47	0.59
Cuntan	Ammonia Nitrogen (mg/L)	12	0.15	0.20	0.05	44.58	0.07	0.08
Qingxichang	Ammonia Nitrogen (mg/L)	12	0.17	0.18	0.01	42.56	0.07	0.09

Station ID	Parameter	Pairs	Data Avg.	Model Avg.	AE	RE	AAE	RMS
Wanzhou	Ammonia Nitrogen (mg/L)	12	0.19	0.15	−0.04	35.93	0.07	0.10
Cuntan	Nitrate Nitrogen (mg/L)	12	1.34	1.25	−0.09	7.85	0.11	0.15
Qingxichang	Nitrate Nitrogen (mg/L)	12	1.48	1.31	−0.17	11.99	0.18	0.21
Wanzhou	Nitrate Nitrogen (mg/L)	12	1.47	1.31	−0.15	12.06	0.18	0.22
Cuntan	Total N (mg/L)	12	1.84	1.90	0.06	11.68	0.22	0.26
Qingxichang	Total N (mg/L)	12	1.90	1.95	0.05	13.85	0.26	0.30
Wanzhou	Total N (mg/L)	12	1.72	1.89	0.17	14.27	0.25	0.31
Cuntan	Chemical Oxygen (mg/L)	12	11.80	12.4	0.61	8.96	1.06	1.60
Qingxichang	Chemical Oxygen (mg/L)	12	12.11	11.79	−0.32	9.61	1.16	1.88
Wanzhou	Chemical Oxygen (mg/L)	12	9.80	10.81	1.01	14.64	1.44	1.65
Cuntan	Total P (mg/L)	12	0.25	0.21	−0.03	24.34	0.06	0.10
Qingxichang	Total P (mg/L)	12	0.29	0.21	−0.09	32.33	0.09	0.18
Wanzhou	Total P (mg/L)	12	0.15	0.22	0.06	61.41	0.09	0.13

The sediment concentration in the Wanzhou Station fluctuated, ranging from <5 mg/L in the low flow period to about 1800 mg/L or more in the high flow period (Fig. 6.4c), so it is quite difficult to simulate sediment process in this station. In this study, the high sediment concentration process is considered mainly to reduce global error (i.e. low AE, RE, AAE and RMS of the sediment simulation). From Table 6.2, model accuracy is higher in the upstream stations due to fewer effects of erosion, settlement, sedimentation and resuspension in the sediment transportation process along the Yangtze River. In Table 6.2, AAE and RMS of sediment are higher than the other variables with a higher concentration value than the others. The same conclusion can be drawn from the simulation results of the temperature (Fig. 6.4d).

Generally, the modeling of DO (Fig. 6.4e) and the nitrogen cycle (Figs. 6.4f, 6.4g, and 6.4i) is better than the others and there are gaps between the simulation results and the measured data of TP (Fig. 6.4j). DO concentration is important for supporting the aquatic ecosystem, and prolonged exposure to less than 60% oxygen saturation may result in altered behavior, growth reduction, adverse reproductive effects and mortality. Fish will begin to feel stress when DO drops to about 4 mg/L and swim away from the area where DO is below 3 mg/L. Mortality of fish will occur and shellfish will begin to shut down (Karim et al., 2002). Sources and sinks of DO in the water column included in the model are: algal photosynthesis and respiration, nitrification, heterotrophic respiration of dissolved organic carbon, oxidation of COD, surface reaeration for the surface layer only, sediment oxygen demand (SOD) for the bottom layer only, and external loads (Park et al., 2000). This study shows that the DO concentration in the water column is primarily controlled by reaction rates of NH_4-N and COD; the DO comparison plot in Fig. 6.4e indicates that the measured DO concentration change is reduplicated by the model by selecting available parameters for NH_4-N and COD reaction.

In Fig. 6.4f, the measured COD was in the range of about 2 mg/L to more than 12 mg/L. The simulation results are larger than the corresponding measured data in the whole year, but both have a same change tendency. The REs (see Table 6.2) in the three stations are as small as 8.96%, 9.61%, and 14.64%. AE, AAE and RMS are also acceptable.

The temporal NH_4-N concentration change is also represented in the model with a maximum nitrification rate of 0.1 g N/(m^3·day) (Fig. 6.4g). According to Table 6.2, AEs are very small in the three stations due to low concentration in the reservoir region. The REs are 35.93% at the Wanzhou Station and 44.58% at the Cuntan Station, which could be due to enormous variation in the field measured data from <0.1 mg/L in the low flow period to 0.45 mg/L in the high flow period. NO_3-N (Fig. 6.4h) and TN (Fig.6.4i) have the same trend as shown in Fig. 6.5, because NO_3-N accounts for more than 75% of TN in the TGR. The REs for NO_3-N and TN simulation are small according to Table 6.2. This can be explained by the fact that the NO_3-N concentration is less affected by external loads, especially by non-point pollution sources during the high flow period.

The model result for TP (Fig.6.4j) before July is better than that for the remaining days, when the measured data are lower than the simulation results. The statistical results in Table 6.2 show that the errors are larger than the results of other variables, in which the RE in WZ is 61%. In the EFDC model, TP is the sum of the phosphorus dissolved in the water and adsorbed on the sediment or active metals (Craig, 2009), but in the laboratory manual in China (Zhen, 2002), TP is tested after a 30min stewing using the supernatant fluid containing part partial sediment. There is no obvious discrepancy when the sediment content in water is low but there is big difference in the Yangtze River with a sediment concentration of about 2000 mg/L in the high flow period; a large portion of TP adsorbed in cohesive sediment particles is ignored when using the Chinese standard analysis method.

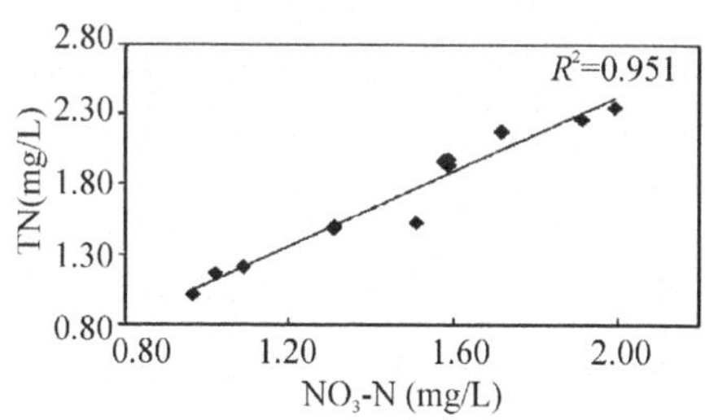

Fig. 6.5 Measured TN vs. NO_3-N curve in Wanzhou station in 2004

● *Longitudinal Profiles Comparison*

To evaluate the established EFDC model in the whole reservoir, longitudinal profiles were plotted and the measured data and simulation results were compared in all 9 stations along the Yangtze River in the TGRR. Results over three days representing the low flow period (02/12/2004), the mean flow period (05/10/2004) and the high flow period (08/08/2004) are given in Fig. 6.6. Water temperature along the river was not plotted because of little spatial variation in the reservoir.

Fig. 6.6a indicates that the water surface elevation is simulated satisfyingly along the river in three flow periods. There are 3 only stations conducting sediment sample analysis. Sediment distribution in the low period is reproduced well by the model (Fig. 6.6b) due to the small variation in sediment contents from upstream. However, sometimes the sediment simulation results from the other two flow periods do not coincide with the measured data. The reason is the frequent occurrence of flooding with high sediment levels. The measured data greatly fluctuate during different sampling times in a flood period.

The simulated DO concentration matches the measured DO data in the upper section in each flow period (Fig. 6.6c), but it is lower than the measured DO in the downstream stations near the TGD. During the field measurement of DO, only surface DO concentration is measured; however, in the model, the simulation DO is the average value at average water depth.

For COD (Fig. 6.6d), the simulation results in the average flow period are better than in the other two periods. There is a sharp increase in COD at a distance of 140 km to 160 km, due to industrial and domestic load emissions in Chongqing City. A large decrease is observed at the confluence with the Jialingjiang River; then the COD level decays till the location of Changshou District (214 km) where an increase in COD concentration appears with the merger of the Wujiang River (258 km). Then, COD decreases again. It increases in Wanzhou District (458 km).

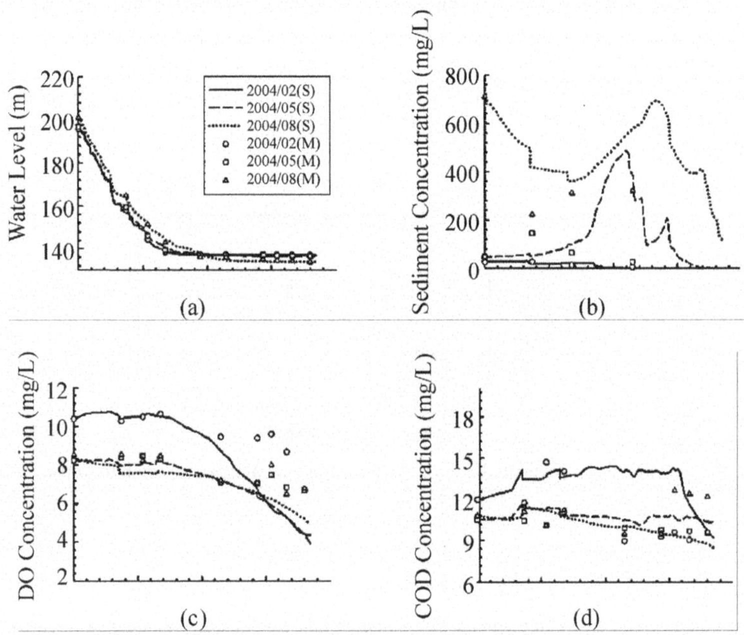

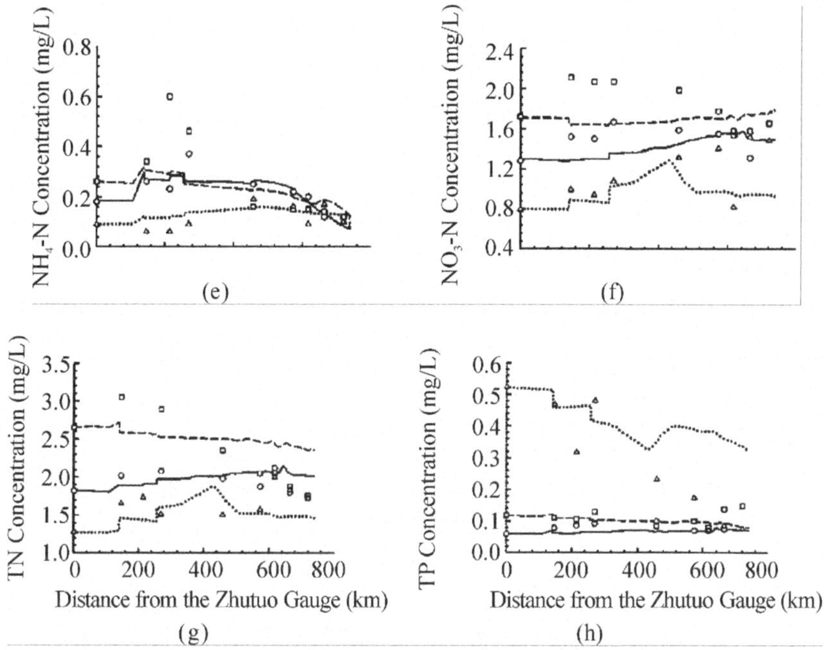

Fig. 6.6 Comparison of measured and simulation results along the Yangtze River, 2004

A similar trend can be seen from the results of NH_4-N, too (Fig. 6.6e).

Generally, NO_3-N (Fig. 6.6f) and TN (Fig. 6.6g) has the same simulation efficacy, though an increase occurs in TN in Chongqing City due to a large discharge of NH_4-N. In contrast to the result in the high flow period, there is a high consistency between the modeled and the measured TP concentration during the low and mean flow period shown in Fig. 6.6h. The measured concentration is lower than that predicted during the high flow period.

6.3.5 Model Validation

A method similar to calibration was employed for the verification process. Measured data from Cuntan, Qingxichang and Wanzhou Stations were used for time series comparison, and the period lasted for 365 days, between January 1 and December 31, 2006. For simplicity, only the results of Wanzhou Station are given in Fig. 6.7. Table 6.3 provides statistical results. Fig. 6.8 shows longitudinal comparison results of hydrodynamic and water quality variables at 9 stations in the whole reservoir region during low, mean and high flow periods.

● *Time series Comparison*

During the high flow period, the measured water surface elevation data (Fig. 6.7a) were higher than the simulated ones, but were close to simulation results in the low and average flow periods. AE, RE, AAE, and RMS are small at the three monitoring stations and the relative errors (RE) are less than 1%. The simulated flow rate data fit with the field measured data except in the high flow period (Fig. 6.7b).

The dynamic changeable process of sediment (Fig. 6.7c) was simulated in the model at Wanzhou Station. There is little difference between the modeled sediment concentration and measured field data at Wanzhou Station, and the AE, RE, AAE and RMS are 17.82, 49.12%, 45.70, and 99.43 (Table 6.3). The reason may be caused by high concentration values during the high flow period not being captured in the model. The high concentration sediment is chiefly from soil erosion upstream in the Chanjing River. Statistical results in Table 6.3 show that the model simulated the sediment levels at the upper stations—Cuntan and Qingxichang—better than at Wanzhou Station (AE, RE, AAE, and RMS are –48.90, 22.16%, 58.39, and 23.23 at Cuntan Station, and are –35.80, 28.05%, 58.59, and 21.99 at Qingxichang Station) in Table 6.3. The simulated temperature is lower than the measured values during the low and average flow periods.

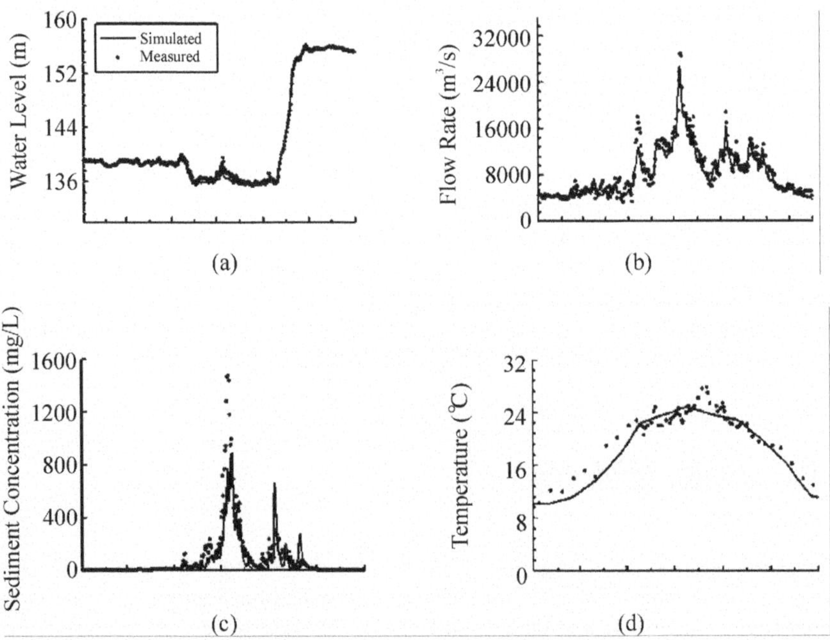

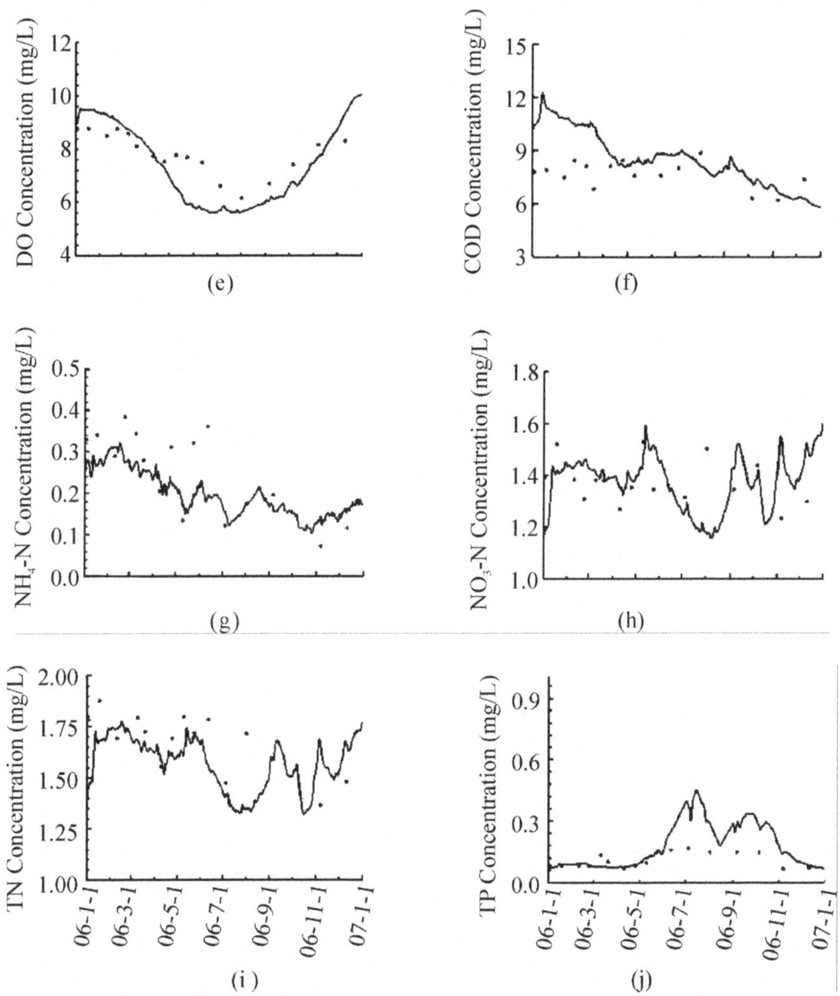

Fig. 6.7 Comparison of simulation results and measured data at Wanzhou Station, 2006

Table 6.3 Simulated results analysis at Cuntan, Qingxichang and Wanzhou Stations, 2006

Station ID	Parameter	Pairs	Data Avg.	Model Avg.	AE	RE	AAE	RMS
Cuntan	Water Surface (m)	365	160.57	160.44	−0.14	0.63	1.01	8.31
Qingxichang	Water Surface (m)	365	142.98	143.89	0.91	0.67	0.96	8.08
Wanzhou	Water Surface (m)	365	140.20	140.09	−0.11	0.14	0.20	0.30
Cuntan	Sediment (mg/L)	365	263.50	214.59	−48.90	22.16	58.39	23.23
Qingxichang	Sediment (mg/L)	365	208.85	173.05	−35.80	28.05	58.59	21.99

Station ID	Parameter	Pairs	Data Avg.	Model Avg.	AE	RE	AAE	RMS
Wanzhou	Sediment (mg/L)	365	93.05	110.86	17.82	49.12	45.70	99.43
Cuntan	Temperature (°C)	56	21.25	20.98	−0.28	4.96	1.05	1.36
Qingxichang	Temperature (°C)	54	21.39	20.65	−0.74	6.01	1.29	1.61
Wanzhou	Temperature (°C)	65	21.71	21.41	−0.30	5.67	1.23	1.59
Cuntan	Dissolved Oxygen (mg/L)	17	7.42	9.15	1.73	23.30	1.73	1.78
Qingxichang	Dissolved Oxygen (mg/L)	18	7.39	9.21	1.82	24.57	1.82	1.93
Wanzhou	Dissolved Oxygen (mg/L)	17	7.83	8.36	0.53	8.65	0.68	0.78
Cuntan	Ammonia Nitrogen (mg/L)	17	0.19	0.17	−0.02	24.48	0.05	0.07
Qingxichang	Ammonia Nitrogen (mg/L)	18	0.15	0.16	0.01	17.56	0.03	0.04
Wanzhou	Ammonia Nitrogen (mg/L)	17	0.24	0.21	−0.03	24.64	0.06	0.07
Cuntan	Nitrate Nitrogen (mg/L)	17	1.26	1.27	0.01	9.67	0.12	0.16
Qingxichang	Nitrate Nitrogen (mg/L)	18	1.33	1.35	0.02	13.05	0.17	0.26
Wanzhou	Nitrate Nitrogen (mg/L)	17	1.37	1.39	0.02	8.87	0.12	0.15
Cuntan	Total N (mg/L)	17	1.75	1.66	−0.10	13.55	0.24	0.42
Qingxichang	Total N (mg/L)	18	1.78	1.74	−0.04	9.22	0.16	0.19
Wanzhou	Total N (mg/L)	17	1.67	1.83	0.17	11.33	0.19	0.23
Cuntan	Chemical Oxygen (mg/L)	17	9.73	10.51	0.78	9.69	0.94	1.24
Qingxichang	Chemical Oxygen (mg/L)	18	9.73	10.35	0.62	11.02	1.07	1.29
Wanzhou	Chemical Oxygen (mg/L)	17	7.72	9.45	1.73	23.23	1.79	2.27
Cuntan	Total P (mg/L)	19	0.15	0.16	0.01	34.17	0.05	0.08
Qingxichang	Total P (mg/L)	18	0.17	0.14	−0.03	21.76	0.04	0.05
Wanzhou	Total P (mg/L)	17	0.11	0.15	0.04	47.59	0.05	0.08

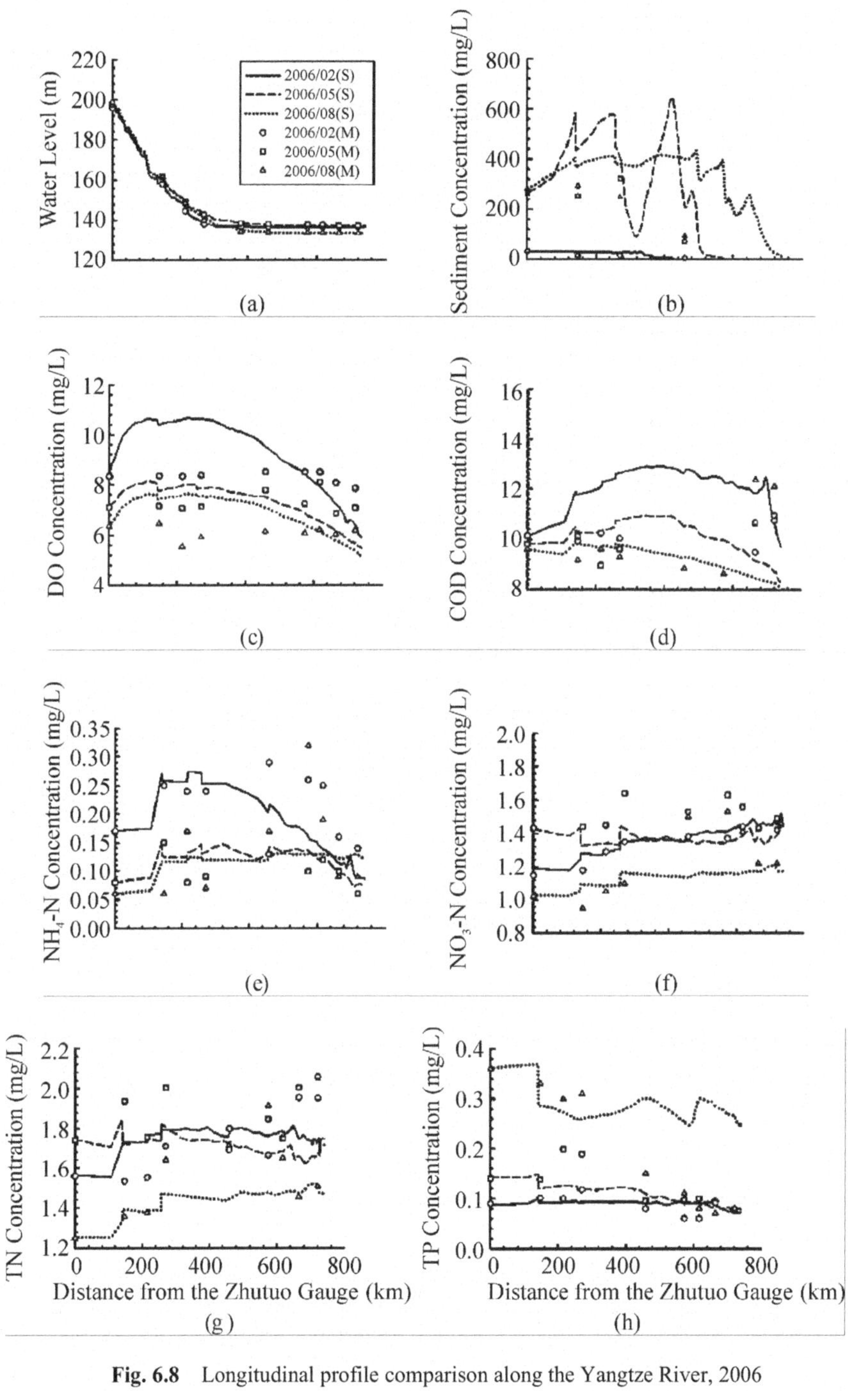

Fig. 6.8 Longitudinal profile comparison along the Yangtze River, 2006

Fig. 6.7e shows that the profile of DO concentrations is simulated with the EFDC model satisfactorily. According to Table 6.3, REs are small at the 3 stations, from 8.65% at Wanzhou Station to 23.30% at Cuntan Station.

Fig. 6.7f illustrates that the simulated COD results are higher than the measured data during the period January to May, 2006, after which the model worked very well. The overall RE is around 25%. In Fig. 6.7g, a part of measured NH_4-N concentration values is higher during the first half of 2006, and the same condition appears in the TN simulation (Fig. 6.7i). There is a high fit between the measured and simulated NO_3-N concentration during 2006 (Fig. 6.7h). Similar to the condition in 2004, the simulated TP concentration data are higher than the measured during the high flow period, as shown in Fig. 6.7j.

The RE of TP is the highest among the simulated variables with a high value of 47.59% at Wanzhou Station, which is the largest among variables because of a different sampling method.

- *Longitudinal Profiles Comparison*

Fig. 6.8a indicates that the water surface elevation was simulated satisfyingly along the river in three flow periods. There are 3 only stations conducting sediment sampling in the reservoir region. Sediment distribution in the low period is reproduced well in the model. However, the sediment model results in the mean and high flow periods do not coincide with the partially measured results (Fig. 6.8b). In Fig. 6.8c, there is a good fitness for the modeled and the measured DO in the low and mean flow periods, but a few differences appear in the high flow period and the results are lower than the measured ones near the Three Gorges Dam. Waste load discharges and decay conditions of COD and NH_4-N along the Yangtze River are revealed in Figs. 6.8d and 6.8e. The model also simulates NO_3-N and TN variation in the water column satisfactorily, illustrated in Fig. 6.8f and Fig. 6.8g. In contrast to the result in the high flow period, there is a high consistency between the modeled and the measured TP concentration during the low and mean flow period shown in Fig. 6.8h.

6.4 Three-Dimensional Eutrophication Modeling at the Daning River Confluence at Mouth of the Three Gorges Reservoir Area

6.4.1 Data Acquisition and Preprocessing

The hydrological, meteorological and water quality data required for the model simulation mainly were collected from the Hydrological Bureau of Changjiang Water Resources Commission, Chongqing Municipal Meteorological Bureau and Wushan County Environmental Monitoring Station, etc. The hydrological data includes water level and flow data at Wuxi and Dachang hydrological cross section, and water level data at Daning River confluence. The meteorological data is comprised of atmospheric pressure, air temperature, relative humidity, precipitation, evaporation, solar radiation, wind speed and wind direction, which are primarily used to configure the asper.inp file. The water quality data includes water quality monitoring data for the upper, middle and lower layer at Dachang, Shuanglong and Longmen hydrological cross sections. The detailed data is comprised of water temperature, COD_{Mn}, TN, TP, Chl-a, DO, ammonia nitrogen and nitrate nitrogen. Besides some water quality variables which can be added into the Daning River model directly (such as DO and water temperature), the other variables are obtained in accordance with the TN and TP proportion with a similar approach to the Three Gorges Model as in 6.3. The variables are divided into cyanobacteria, green algae and diatoms according to the proportion of Chl-a and algae.

6.4.2 Model Configuration

The study area was subdivided into 1560 cells horizontally and 4 layers vertically by using Delft3D software, and these cells were imported into the EFDC_Explorer, so each layer had 1560 cells and 6240 cells in total (Fig. 6.9). The calibration period was from February 10, 2006 to November 11, 2006, and the validation period was from January 24, 2007 to October 1, 2007 according to the data availability. The dynamic time step was adopted, and the minimum step length was set to 0.5 s. The computing time for once need approximately 40 h (AMD Dual Core 1300+, 1 GB memory). Setting the Dachang hydrological cross section as flow boundary, its flow data was calculated from data at Wuxi hydrological station. The river mouth was set as an open boundary condition. The calculation adopted an improved EFDC.exe file, which improved the algae reaction module and added a module reflecting the hydrodynamic effects on the algae (Eqs. (6.15) to (6.17)).

The original equation:

$$Px = PMx \times f(N) \times f(I) \times f(T) \tag{6.15}$$

The improved equation:

$$Px = PMx \times f(N) \times f(I) \times f(T) \times f(V) \tag{6.16}$$

where

$$f(V) = 0.7^{6.6V} \tag{6.17}$$

where P_x is production rate of algal group x; PM_x is maximum growth rate under optimal conditions for algal group x; $f(N)$ is effect of suboptimal nutrient concentration, $0<f(N)<1$; $f(I)$ is effect of suboptimal light intensity, $0<f(I)<1$; $f(T)$ is effect of suboptimal temperature, $0<f(T)<1$; V is flow velocity.

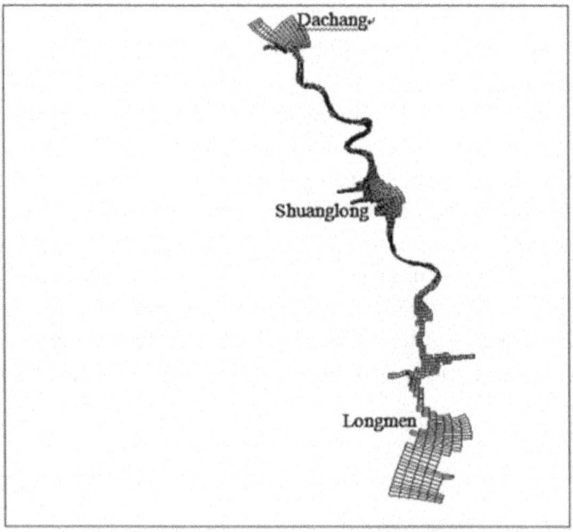

Fig. 6.9 Orthogonal curvilinear grids of the Daning River eutrophication model

6.4.3 Parameter Estimation

Used parameters are listed in Table 6.4.

6.4.4 Model Calibration

The model calibration aims to reduce the error between model output and measured value by adjusting the parameter values. The calibration for the Daning River model was carried out by manual operation and calculated approximately 150 times. As the study area is in the backwater zone, the height of water variance is small and the hydrodynamic process is unobvious. In addition, there are no hydrological stations in the study area. Therefore, this study did not carry out the simulation work of the water level at each hydrological cross section and only provided flow velocity simulation results from the Dachang stream segment to the Daning River confluence (Fig. 6.10), which aimed to analyze the changes in water quality condition. The model calibration was carried out by comparing the simulated and measured values at the Shuanglong and Longmen hydrological cross section. The Shuanglong and Longmen hydrological cross section are routine environmental monitoring sections of Wushan Environmental Monitoring Station. Each section has upper, intermediate and lower monitoring points, which can supply support for the calibration of the three-dimensional model.

According to Fig. 6.10, the flow field of the Daning River backwater zone is controlled by influx from upstream Dachang segment and return water from the main stream. The flow direction is not unidirectional like ordinary flow direction. In the wet season, the amount of water from upstream is large and the water flows into the Yangtze River through the Daning River confluence. The Three Gorges Reservoir begins to store water and the water flows backward from the main stream to the backward zone after the flood season.

Table 6.4 Parameters of the Daning River eutrophication model

Parameter	Unit	Value
Roughness height	m	0.02
Specific volume	m^3/g	3.77E-07
Specific gravity	-	2.65
Settling velocity	m/s	0.00002
Tau critical-deposition	m^2/s^2	0.001
Tau critical-erosion	m^2/s^2	0.002
Ref. surf erosion rate	$g/(m^2 \cdot s)$	0.0005
Maximum nitrification rate	$g\ N/(m^3 \cdot day)$	0.0005
Oxygen half-sat constant for nitrification	$g\ O_2/m^3$	0.1
NH_4 half-sat constant for nitrification	$g\ N/m^3$	0.1
Reference temperature for nitrification		20
Partition coefficient for sorbed/dissolved PO4		0.04
Settling velocity for refractory particulate organic matter	m/day	1
Settling velocity for liable particulate organic matter	m/day	1
Reaeration constant		3.933
Temperature rate constant for reaeration		1.024
Maximum growth rate for algae	$1\ day^{-1}$	5.5
Basal metabolism rate for cyanobacteria	$1\ day^{-1}$	0.01
Basal metabolism rate for diatoms	$1\ day^{-1}$	0.05
Basal metabolism rate for greens	$1\ day^{-1}$	0.01
Predation rate on cyanobacteria	$1\ day^{-1}$	0.1
Predation rate on diatoms	$1\ day^{-1}$	0.05
Background light extinction coefficient	$1\ m^{-1}$	0.475
Light extinction due to TSS	$1\ m^{-1}$ per mg/L	0.015
Lower optimal temperature for growth, Cyanobacteria	°C	16
Upper optimal temperature for growth, Cyanobacteria	°C	20
Lower optimal temperature for growth, Diatoms	°C	9
Upper optimal temperature for growth, Diatoms	°C	15
Lower optimal temperature for growth, Greens	°C	10
Upper optimal temperature for growth, Greens	°C	30
Optimal depth for growth	m	1
Settling velocity for Cyanobacteria	m/day	0
Settling velocity for Diatoms	m/day	0.02
Settling velocity for Greens	m/day	0.02

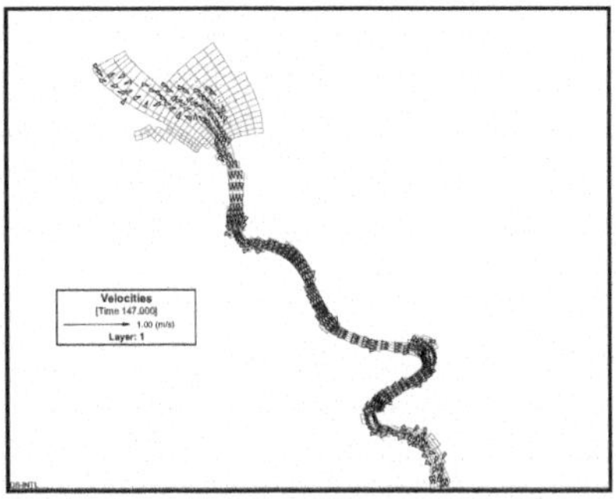

(a) Dachang stream segment

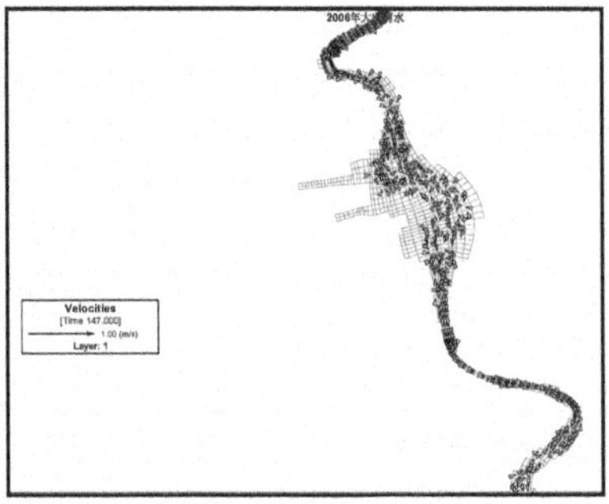

(b) Shuanglong stream segment

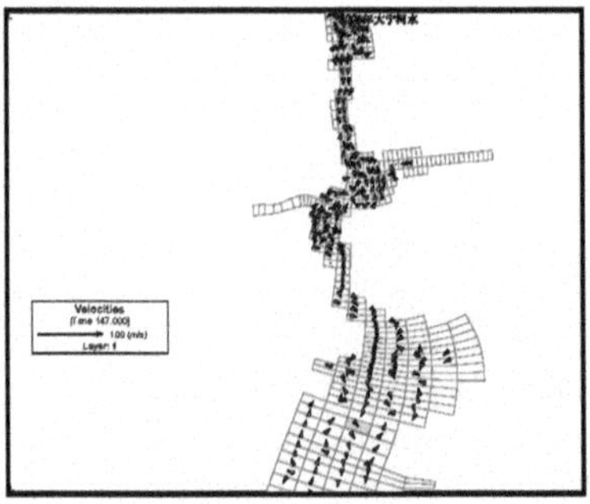

(c) Daning River confluence mouth

Fig. 6.10 Velocity field

The comparison between simulated results and measured values of water temperature, sediment, DO, TN, TP, Chl-a at the Shuanglong Section are shown in Fig. 6.11, where S represents the surface layer, M represents the middle layer, and B represents the lower layer. The statistical results for the water quality variables at the Shuanglong Section are shown in Table 6.5. The comparison between simulated and measured values and the statistical results of water quality variables at the Longmen Section will not be shown here.

(1) Calibration Results at the Shuanglong Section

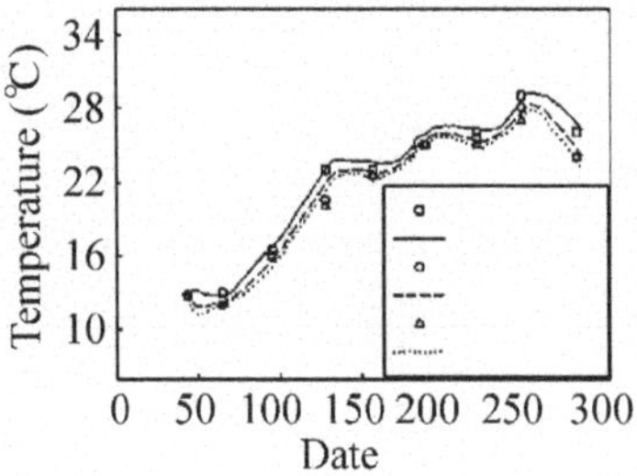

(a) Temperature

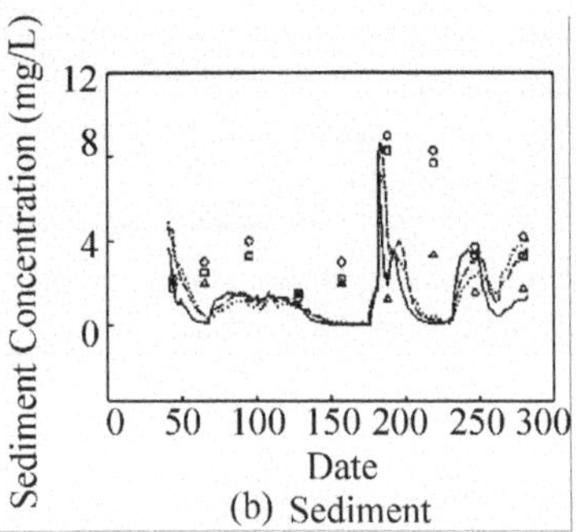

(b) Sediment

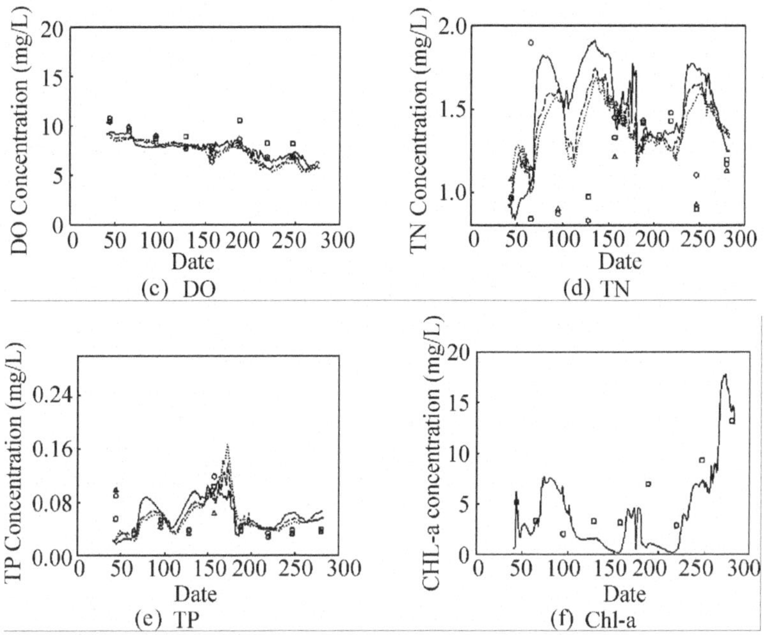

Fig. 6.11 Calibration results at the Shuanglong Section, 2006 (S-upper layer; M-intermediate layer; B-lower layer)

Table 6.5 Statistical results of water variables at the Shuanglong Section, 2006

Variable	Layer	Measured value	Calculated value	AE	RE	AAE	RMSE
Dissolved oxygen (mg/L)	S	9.07	8.03	−1.03	11.40	1.03	1.23
Dissolved oxygen (mg/L)	M	8.20	7.79	−0.41	9.43	0.77	0.86
Dissolved oxygen (mg/L)	B	8.16	7.71	−0.45	8.86	0.72	0.83
Sediment (mg/L)	S	3.88	1.20	−2.67	72.85	2.82	3.65
Sediment (mg/L)	M	4.26	1.69	−2.57	73.44	3.13	3.85
Sediment (mg/L)	B	4.98	1.69	−3.29	79.04	3.94	4.56
Temperature (°C)	S	20.38	21.47	1.10	5.65	1.15	1.54
Temperature (°C)	M	19.31	20.54	1.23	6.87	1.33	1.68
Temperature (°C)	B	18.94	20.10	1.16	7.39	1.40	1.70
Total N (mg/L)	S	1.07	1.47	0.40	39.08	0.42	0.57
Total N (mg/L)	M	1.25	1.36	0.11	31.41	0.39	0.51
Total N (mg/L)	B	1.19	1.33	0.14	33.87	0.40	0.50
Total P (mg/L)	S	0.05	0.05	0.01	43.78	0.02	0.03
Total P (mg/L)	M	0.06	0.05	−0.01	47.55	0.03	0.03
Total P (mg/L)	B	0.05	0.05	0.00	46.49	0.02	0.03
Chl-a (μg/L)	S	4.51	1.26	−3.26	72.19	3.26	4.12

The changes in water temperature affect the reaction rate of each water quality variable. The accurate simulation of water temperature is the prerequisite of high quality simulation of water quality. As can be seen from Fig. 6.11a, the simulated water temperature was in good conformity with measured values at the Shuanglong Section. The values of AE, RE, AAE, RMSE are relatively small. Take the RE as an example; the largest relative error was just 7.93% among the upper, middle and lower layers at the Shuanglong Section.

Although the simulated results of sediment were lower than the measured values (Fig. 6.11b), the entire trend showed no difference, and there were better results in the vicinity of the crest values. It also can be seen from Table 6.5 that the RE was close to 80% because of small simulated results. However, these unfavorable results had little impact on the transparency and algae growth due to low concentration of sediment. Therefore, these values had little impact on the final eutrophication assessment as well.

From Fig. 6.11c, it can be seen that the simulated values of DO concentration were ranged from 5 mg/L to 10 mg/L, which were close to the measured values. As can be seen from Table 6.5, the largest AE, RE, AAE and RMSE of simulated DO values among upper, middle and lower layers were −1.03, 11.40%, 1.03 and 1.23, respectively.

It can be seen from Fig. 6.11d that the simulated results of TN were quite different from measured data. There are probably two reasons for this phenomenon; one is poor simulation results, the other is the big sampling error (there were big variances in measured TN between the upper layer and lower layer). The statistical indicators of TN were not great and the biggest RE for TN indicators was just 39.08% (Table 6.5).

As can be seen from Fig. 6.11e, there was a big difference between simulated and measured TP initially. However, the difference became much smaller as time went on. The reason for these results may be that the deviation between the actual initial situation and setting the initial condition was big, and it gradually decreased as iteration time went by.

Here are only given the comparison results between simulated and measured values of the upper layer because the Wushan environmental monitoring department only analyzes the Chl-a of the upper layer currently. From Fig. 6.11f, it can be seen that the Daning River model can on the whole simulate the change process in Chl-a.

(2) Spatial Distribution of Water Quality Variables

Because the Daning River eutrophication model is a 3-D model, the spatial distribution of water quality variables changes with time variation. For reasons of space and clarity, this study will not provide relative figures. This chapter takes the dynamic variation of Chl-a on a certain three days for example. Figs. 6.12a, 6.12b, 6.12c shows the spatial distribution of Chl-a between July 20 and July 22, 2006. The maximum concentration was 1.831 μg/L and it appeared in the stream segment between Shuanglong and Longmen Sections.

6.4.5 Model Validation

The water quality data collected in the year of 2007 were used to test the effectiveness of calibrated data for the Daning River model. The calibration phase was also conducted at the Shuanglong and Longmen Sections. The simulated results of flow velocity can be seen from Fig. 6.13. Compared with calibrated results, the validated results had a similar trend. The comparison between simulated and measured data at the Longmen Section can be seen from Fig. 6.14, and the statistical results of the validation phase at the Longmen Section are shown in Table 6.6. The results at the Shuanglong Section will not be shown here.

(1) Validation results at the Longmen Section

Similar to the calibration results, the simulation of water temperature was comparatively favorable (Fig. 6.14a). From Table 6.6, it can be seen that the values of AE, RE, AAE and RMSE for temperature simulation were comparatively small and the maximum relative error of upper, middle and lower layers was only 12.85%.

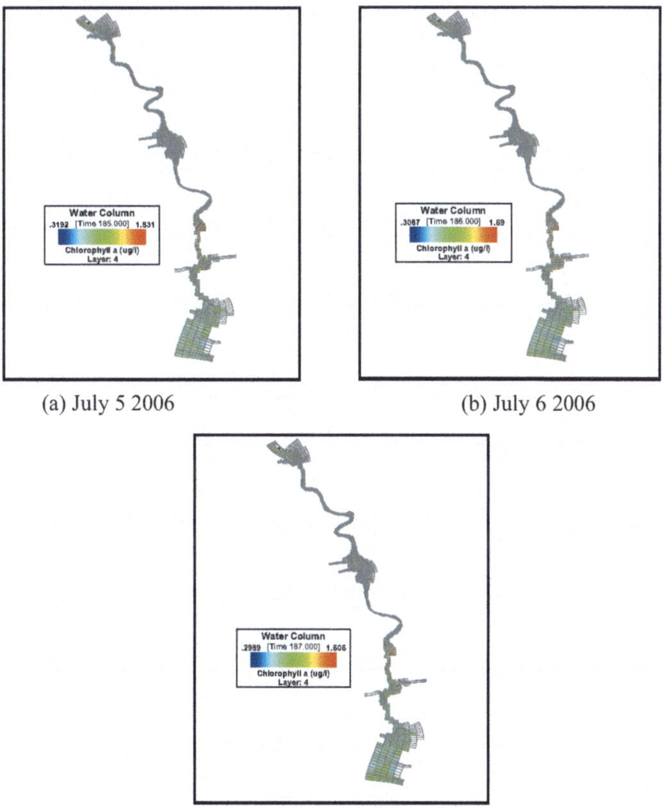

(a) July 5 2006

(b) July 6 2006

(c) July 7 2006

Fig. 6.12 Spatial distribution of Chl-a

The simulated results for sediment had the same tendency as measured values at the Longmen Section (Fig. 6.14b). However, the concentration difference in each layer was larger than the measured values. The data from Table 6.6 indicates that all the simulated results were larger than relative measured values. The maximum value of AE (1.65) appeared in the middle layer, while the maximum value of RE (54.89%) appeared in the upper layer. Both the maximum values of AAE and RMSE appeared in the bottom layer.

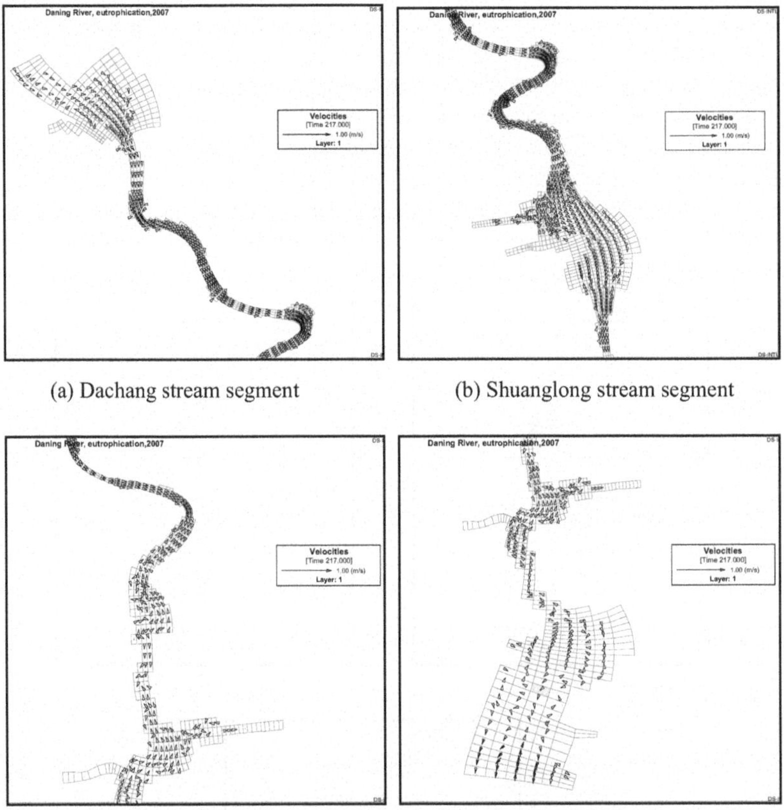

(a) Dachang stream segment (b) Shuanglong stream segment

(c) Pipazhou stream segment (d) Longmen stream segment

Fig. 6.13 Velocity field in the Daning River

As can be seen from Fig. 6.14c, the simulated and measured concentrations of DO ranged from 5 mg/L to 10 mg/L, and both had a similar variation tendency. From Table 6.6, it can be seen that each variable was comparatively small, and the RE was always less than 20%.

It can be seen from Fig. 6.14d that the simulated TN had a similar tendency to measured values during the first half of the time. However, there were some errors in measured and simulated concentration during the latter half-time, when the measured concentration decreased to a certain extent. The statistical data from Table 6.6 indicated that each statistical variable was comparatively small. The maximum values of AE and RE were 0.2 and 21.92%, respectively. And the maximum values of AAE and RMSE were both less than 0.5.

From Fig. 6.14e it can be seen that the Daning River model could simulate the variation tendency of TP basically, and the simulation results were better than those at the Shuanglong Section. The data from Table 6.6 indicated that each variable was comparatively small, and the maximum RE of upper, middle and lower layers was just 40.08%.

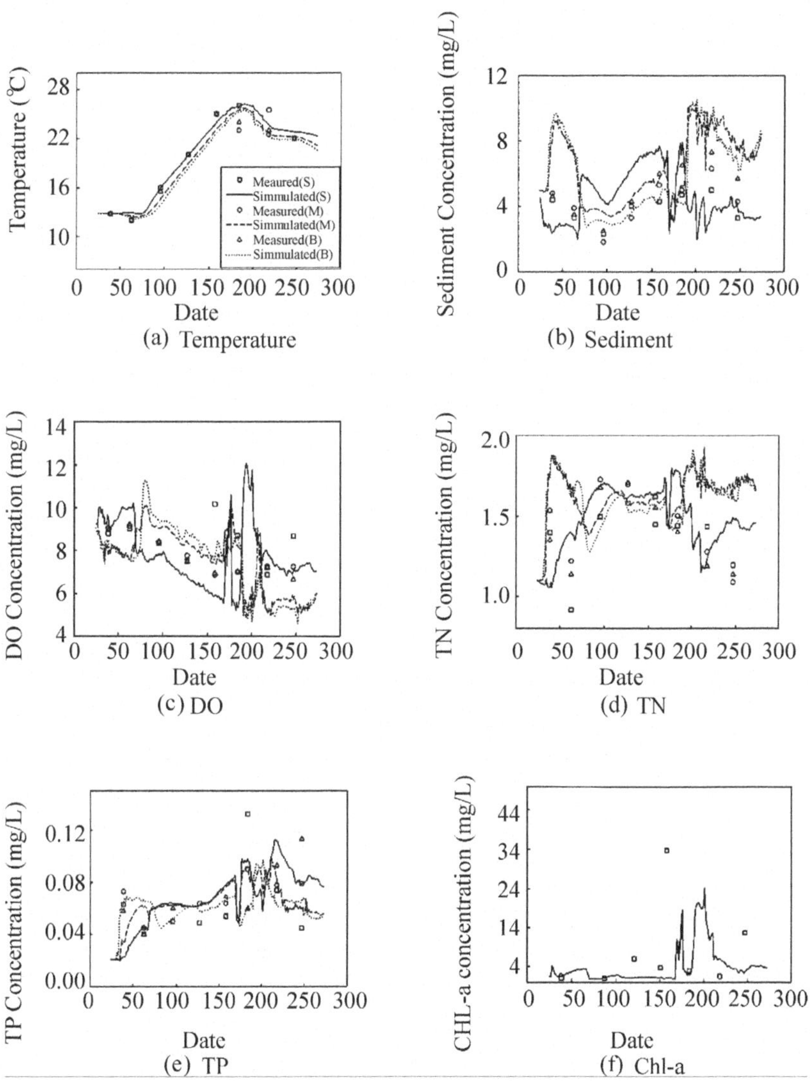

Fig. 6.14 Simulation results at the Longmen Section, 2007 (S-upper layer; M-intermediate layer; B-lower layer)

It can be seen from Fig. 6.14f that, in general, the Daning River model could simulate the variation process but not the crest value of Chl-a. Moreover, the relative error was comparatively large, with a value of 93.03%.

Table 6.6 Statistical results of water quality variables at the Longmen section, 2007

Variable	Layer	Measured value	Calculated value	AE	RE	AAE	RMSE
Dissolved oxygen (mg/L)	S	8.51	7.55	–0.97	19.27	1.64	2.10
Dissolved oxygen (mg/L)	M	7.83	7.41	–0.43	15.18	1.19	1.30
Dissolved oxygen (mg/L)	B	7.73	7.66	–0.07	16.53	1.28	1.37
Sediment (mg/L)	S	3.84	4.58	0.74	54.89	2.11	2.72
Sediment (mg/L)	M	4.94	6.59	1.65	51.48	2.55	3.14
Sediment (mg/L)	B	5.47	6.00	0.53	48.52	2.65	3.18
Temperature (°C)	S	20.50	19.71	–0.79	9.37	1.92	2.55
Temperature (°C)	M	19.81	18.80	–1.01	10.80	2.14	2.85
Temperature (°C)	B	19.44	18.44	–1.00	12.85	2.50	3.10
Total N (mg/L)	S	1.36	1.49	0.13	18.99	0.26	0.29
Total N (mg/L)	M	1.46	1.64	0.18	17.97	0.26	0.33
Total N (mg/L)	B	1.40	1.59	0.20	21.92	0.31	0.36
Total P (mg/L)	S	0.06	0.07	0.01	40.08	0.03	0.03
Total P (mg/L)	M	0.07	0.06	0.00	18.21	0.01	0.01
Total P (mg/L)	B	0.06	0.06	0.00	33.09	0.02	0.03
Chl-a (μg/L)	S	7.82	2.55	–5.27	93.03	7.28	12.31

(2) Spatial Distribution of water quality variables

The spatial distribution of Chl-a between July 5 and July 7, 2007, can be found in Fig. 6.15. The maximum concentration was 44.59 μg/L. Consistent with the calibration phase, the maximum concentration appeared in the stream segment between the Shuanglong and Longmen Sections.

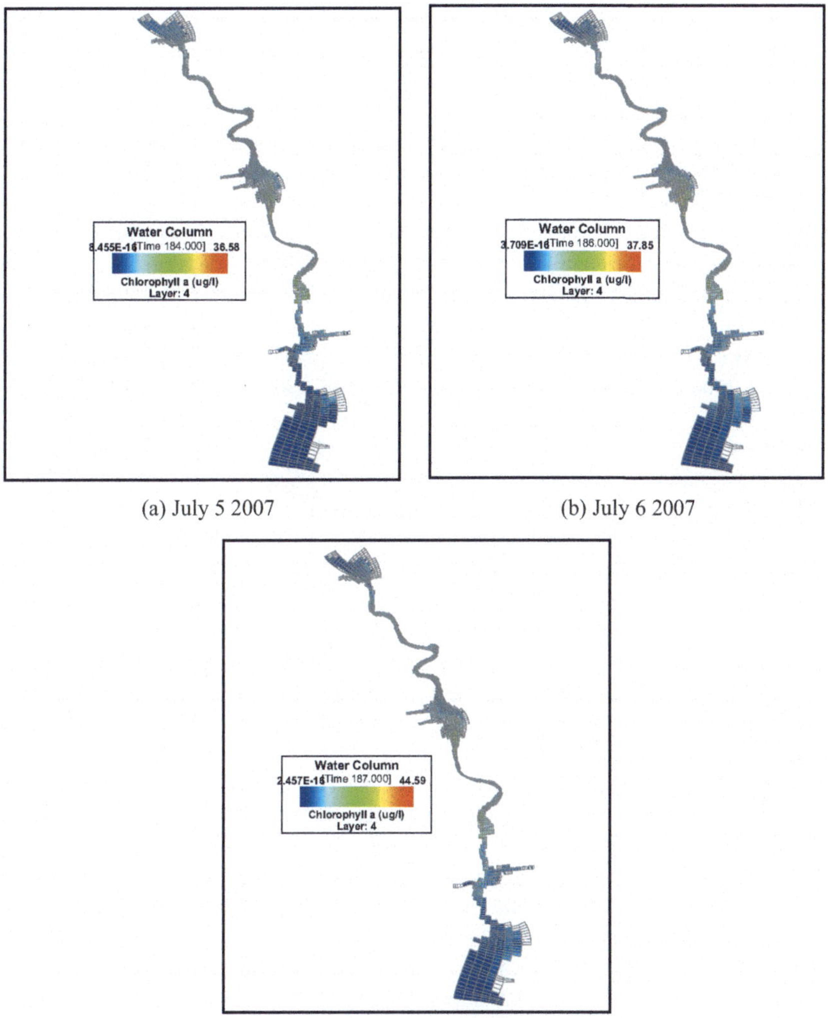

(a) July 5 2007

(b) July 6 2007

(c) July 7 2007

Fig. 6.15 Spatial distribution of Chl-a in Daning River

References

Cao YL (2006) Nonpoint pollutant source analysis and its numerical simulation study in Chongqing, Three Gorges Reservoir Region. M.S. Thesis, Chongqing University, Chongqing.

Cerco CF, Cole TM (1993) Three-dimensional eutrophication model of

Chesapeake Bay. J. Environ. Eng. 119:1006-1025.

Cerco CF, Cole TM (1994) Three-dimensional eutrophication model of Chesapeake Bay; Vol. 1, Main report, Techn. Report. EL-94-4, U.S. Army Engineer Waterways Experiment Station, Vicksburg, M.S.

Chapra SC, Pelletier GJ, Tao H (2007) QUAL2K: A Modeling. Framework for Simulating River and Stream Water Quality, Version 2.07: Documentation and Users Manual. Civil and Environmental Engineering Dept., Tufts University, Medford, MA.

Cole TM, Buchak E (1995) CE-QUAL-W2: A Two-dimensional, Laterally Averaged, Hydrodynamics and Water Quality Model, Version 2.0. U.S. Army Engineers Waterways Experiment Station, Tech. Report EL-95-May 1995, Vicksburg, M.S.

Craig PM (2009) User's Manual for EFDC_Explorer: A Pre/Post Processor for the Environmental Fluid Dynamics Code. Dynamic Solutions Intl, LLC, Knoxville, TN.

CWRPI, CAS (1991) Report on the impact of the Three Gorges Project on the ecology and environment, Changjiang Water Resources Protection Institute,Chinese Academy of Sciences, Wuhan.

Deng CG (2007) The study of eutrophication in the Three Gorges Reservoir Region. Chinese Environment Press, Beijing, China, 197.

Denham CR (2006) SeaGrid Orthogonal Grid Maker for Matlab. U.S. Geological Survey, 384 Woods Hole Road, Woods Hole, Massachusetts 02543. http://woodshole.er.usgs.gov/staffpages/cdenham/public_html/seagrid/seagrid.html.

DHI (1995) MIKE 21 short description, Danish Hydraulic Institute, Hørsholm, DenmarkDHI, 1999. MIKE SHE Water Movement: User manual, Danish Hydraulic Institute, Hørsholm, Denmark.

Di Toro DM, Fitzpatrick J (1993) Chesapeake Bay sediment flux model Contract Report EL-93-2, U.S. Army Engineer Waterways Experiment Station, Vicksburg, MS.

Dillon PJ (1975) The phosphorus budget of Cameron Lake. Ontario: the importance of flushing rate to the degree of eutrophy of lakes. Limnol Oceanogr 20: 28-39.

Edmonds RL (1992) The Sanxia (Three Gorges) Project: The environmental argument surrounding China's super dam. Glob. Ecol. Biogeogr. 2:105-125.

Galperin B, Kantha LH, Hassid S, Rosati A (1988) A Quasi-equilibrium turbulent energy model or geophysical flows. J. Atmos. Sci. 45:55-62.

George BA, Michael TB (2005) Eutrophication model for Lake Washington (USA). Ecol Model 187:179-200.

Hamrick JM (1992) A Three-Dimensional Environmental Fluid Dynamics Computer Code: Theoretical and Computational Aspects, Virginia Institute of Marine Science, Gloucester Point, VA.

Hamrick JM (1996) A User's Manual for the Environmental Fluid Dynamics Computer Code (EFDC). Virginia Institute of Marine Science, Gloucester Point, VA.

Henderson T, Schaffer D, Govett M, Middlecoff A, Hart L (2004) SMS user's guide, Advanced Computing Branch Aviation Division NOAA/Forecast

Systems Laboratory Boulder, Colorado.
Huang ZL, Li YL, Chen YC, Li JX, Xing ZG (2006) Water quality prediction and water environmental carrying capacity calculation for Three Gorges Reservoir. Beijing: China Waterpower Publisher.
Imboden DM (1974) Phosphorus model of lake eutrophication. Limnol Oceanogr 19:297-304.
Imboden DM, Gachter RA (1978) Dynamic lake model for trophic state prediction. Ecol Model 4:77-98.
Ji ZG (2001) Wetting and drying simulation of estuarine processes. Estuar Coast. Shelf Sci 53:683-700.
Ji ZG (2008) Hydrodynamics and water quality: modeling rivers, lakes and estuaries Wiley-Interscience, Hoboken, New Jersey, 676.
Jørgensen SE, Ray S, Berec BL, Straskraba M (2002) Improved calibration of a eutrophication model by use of the size variation due to succession. Ecol Model 153:269-277.
Karim MR, Sekine M, Ukita M (2002) Simulation of eutrophication and associated occurrence of hypoxic and anoxic condition in a coastal bay in Japan. Mar Pollut Bull 45:280-285.
Kim SC, Wright DL, Maa JPY, Shen J (1998) Morphodynamic responses to extratropical meteorological forcing on the Inner shelf of the Middle Atlantic Bight: wind wave, currents, and suspended sediment transport. Estuar Coast Model 456-466.
Koelmans AA, Van Der Heijde A, Knijff LM, Aalderink RH (2001) Integrated modeling of eutrophication and organic contaminant fate & effects in aquatic ecosystems. A review. Wat Res 35(15):3517-3536.
Li HM, Meng FY, Du GS, Liu XD, Liu LJ, Song F (2007) Analysis of the phytoplankton and water quality in eastern and western Miyun Reservoir. J. Lake Sci 19:146-150.
Li J, Liao W, Huang Z (2002) Numerical simulation of water quality for the Three Gorges Project Reservoir. Shuili Xuebao 12:7-10.
Lin J, Kuo AY (2003) A model study of turbidity maxima in the York River Estuary, Virginia. Estuar 26:1269-1280.
Liu HL, Jin XC, et al. (1987) Lake eutrophication investigation criteria. Beijing: China Environmental Science Press.
Liu Y, Chen JN (2006) Substance flow Analysis on phosphorus cycle in Dianchi Basin. China Environ. Sci. 27:1549-1553.
Malmaeus JM, Håkanson L (2004) Development of a Lake Eutrophication model. Ecol Model 171:35-63.
Mellor GL, Yamada T (1982) Development of a turbulence closure model for geophysical fluid problems. Rev Geophys Space Phys 20:851-875.
MEP (2004) Three Gorges Bulletin in 2004, Ministry of Environmental Protection, http://english.mep.gov.cn/standards_reports/threegorges P.R. China, Beijing, China. bulletin/bulletin2004/.
MEP (2005) Three Gorges Bulletin in 2005, Ministry of Environmental Protection, P.R. China, Beijing, China. http://english.mep.gov.cn/standards_reports/three gorgesbulletin/bulletin2005/.

MEP (2006) Three Gorges Bulletin in 2006, Ministry of Environmental Protection, http://english.mep.gov.cn/standards_reports/threegorges P.R. China, Beijing, China. bulletin/bulletin2006/.

MEP (2007) Three Gorges Bulletin in 2007, Ministry of Environmental Protection, http://english.mep.gov.cn/standards_reports/threegorges P.R. China, Beijing. China, bulletin/bulletin2007/.

Park K, Kuo AY (1996) A multi-step computation scheme decoupling kinetic processes from physical transport in water quality models. Water Res 30:2255-2264.

Park K, Shen J, Kuo AY (1998) Application of a multi-step computation scheme to an intratidal estuarine water quality model. Ecol Model 110:281-292.

Park K, Kuo AY, Shen J, Hamrick JM (2000) A three-dimensional hydrodynamic-eutrophication model (HEM-3D): description of water quality and sediment process submodels, Virginia Institute of Marine Science, Gloucester Point, VA.

Peng ZZ, Yang TX, Liang XJ, et al. (2007) Water Environment Mathematical Model and its Application. Chemical Industry Press, Beijing.

Sun YF, Guo HC (2002) Characteristic analysis and control strategies for the eutrophicated problem of Lake Dianchi. Pro Geogr 21:500-506.

Vollenweider RA (1975) Input-output models with special reference to the phosphorus loading concept in limnology. Schweizerische Zeitschrift fur Hydrologie 37:53-831.

Wang Y, Shen ZY, Hu LJ, Niu JF (2008) Adsorption and release of phosphorus from sediments from the main branches of the Three Gorges Reservoir. Acta Sci Circumstantiae 28:1654-1661.

Wool TA, Ambrose RB, Martin JL, Comer EA (2001) Water quality analysis simulation program (WASP) version 6.0 draft: user's manual. US Environmental Protection Agency-Region 4 Atlanta, US.

Wu J, Huang J, Han X, Gao X, He F, Jiang MX, Primack ZJ, Shen Z (2004) The Three Gorges Dam: an ecological perspective. Front. Ecol. Environ. 2:241-248.

Yin FC, Hang ZY (2003) Survey of Lake Chaohu eutrophication research. J Lake Sci 15:377-384.

Yu XZ (2008) Simulation Study on the Effect of Sediment on Phosphorus in the Three Gorges Reservoir. Ph.D Thesis, Beijing Normal University, Beijing.

Zhang JX, Liu ZJ, Sun XX (2009) Changing landscape in the Three Gorges Reservoir Area of Yangtze River from 1977 to 2005: Land use/land cover, vegetation cover changes estimated using multi-source satellite data. Int J Appl Earth Obs 11:403-412.

Zhao X, Shen ZY, Xiong M, Qi J (2011) Key Uncertainty Sources Analysis of Water Quality Model Using the First-Order Error Method. Int J Envir Sci Technol 8(1):137-148.

Zhen B (2002) Monitoring and Test Methods for Water. China Environmental Sciences Press, Beijing.

Zhen B, Wang L, Gong B (2009) Load of non-point source pollutants from upstream rivers into Three Gorges Reservoir. Res Environ 22:125-131.

7

Eutrophication Risk Assessment

7.1 Overview

Eutrophication is commonly considered as one major aspect of global environmental degradation (Nixon, 1995). To protect and manage the water quality, indices are useful tools to communicate with managers because they reduce complex scientific data, integrate different types of information and produce results that can be easily interpreted in the perspective of water quality management (Potapova and Charles, 2007). Many researchers consider that the main reason for the dramatic propagation of algae is the increase in concentration of nutrient materials, such as nitrogen, phosphate, etc. Nonetheless, the relationships between physiochemical and biological factors that impact the growth of algae, such as sunlight, nutrient salts, changes of seasons, water temperature, pH value, and the algae itself, are very complex. It is very hard to forecast the trend in algal growth and to set the eutrophication indicator. A common method was to indicate the water nutrient level using the main representative parameters. The parameters mainly include the concentrations of total phosphate (TP), and total nitrogen (TN), transparency, concentration of chlorophyll a, dissolved oxygen (DO), and so on. According to the parameters, the lakes are divided into different trophic levels, such as poor, middle, eutrophic and so on. The trophic level criterion is different for different countries and different reseachers. Tables 7.1, 7.2 and 7.3 show the trophic level proposed by the United States Environmental Protection Agency and used by Dianchi Lake and East Lake in Wuhan.

There also are some other eutrophication standard parameters, such as primary productivity of phytoplankton (Catherine et al., 2008), biomass of phytoplankton (Ho et al., 2008), cell density or numbers of phytoplankton (Dorte et al., 2008), community structure and dominant species (Lorenzo et al., 2008), category composition and dominant species of zooplankton (Toru et al., 2008), numbers of zooplankton, category composition and dominant species of zoobenthos, numbers of bacteria, etc. For different monoids, the methods and standards proposed by different scientists are different. These bio-indicators are not sensitive to changes in aquatic environments and thus are unable to reflect the eutrophic conditions of freshwater immediately (Simboura and Zenetos, 2002). To date, there is no consensus among managers and scientists concerning which of the numerous existing indices should be used (Chainho et al., 2007).

Table 7.1 Environmental protection agency's eutrophication standard

Indicator	Poor	Middle	Eutrophication
TP (mg/L)	<0.01	0.01–0.02	>0.02
Chl-a (μg/L)	<4	4–10	>10
Transparency (m)	>3.7	2.0–3.7	<2.0
DO in deep water (saturation%)	>80	10–80	<10

Table 7.2 Chinese Dianchi Lake's eutrophication standard

Indicator	Poor	Middle	Middle eutrophication	Eutrophication	Heavy eutrophication	Unusual eutrophication
Chl-a (mg/m^3)	5	10	15	25	100	>100
TP (mg/L)	0.01	0.025	0.05	0.1	0.5	>0.5
TN (mg/L)	0.12	0.30	0.60	1.20	6.00	>6.0
Transparency (m)	2.0	1.5	1.0	0.7	0.4	<0.1
BOD_5 (mg/L)	1.5	2.0	3.0	5.0	15	>15
COD (mg/L)	2.0	3.0	4,0	7.0	20	>20

Table 7.3 Chinese eutrophication standard of East Lake in Wuhan

Indicator	Poor	Middle earlier stage	Middle later stage	Eutrophication	Heavy eutrophication
TP (mg/L)	<0.1	0.1–0.2	0.2–0.3	0.3–1.0	>1.0
TN (mg/L)	<0.001	0.001–0.005	0.005–0.01	0.01–0.05	>0.05
BOD_5 (mg/L)	<1	1–3	3–5	5–8	>8
COD (mg/L)	1	3	5	8	12
Transparency (m)	>4	4–2	2–1	1–0.5	<0.5

Microorganisms are generally highly sensitive to the surrounding environment and are profoundly alerted by its perturbations. The water quality is strongly influenced by microbial community dynamics and ecosystem functions, such as organic matter contents and nutrient recycling (Celussi and Cataletto, 2007). Eutrophication, in response to hypernutrification by N and P loads, is directly related to the N and P cycling in aquatic environments (Jergensen and Richardson, 1996; Nixon, 1995; Paerl, 1997). All major transformations of N and P in the environment are carried out exclusively by microorganisms including bacteria associated with N: ammonifying bacteria (AB), NFB, AOB, NOB, nitrate-reducing bacteria (NRB), denitrifying bacteria (DNB) (Davidson et al., 2007), and bacteria associated with P: inorganic phosphate-solubilizing bacteria (IPB) and OPB (Kim et al., 2003; Zboinska et al., 1992). It is predicted that the functional bacteria associated with N or P will be potential indicators for monitoring the eutrophic condition.

The Three Gorges Project, the largest dam project in the world, is located in the mid-downstream area of the Yangtze River in Hubei and Sichuan Provinces, China. Despite the benefits of the dam in terms of power generation and flood control, the project has attracted attention for its potential impact on ecosystems and socio-economic stability (Hwang et al., 2007). Especially in recent years, water bloom has occurred frequently in the backwater areas of the TGR in summer and autumn. Proper bio-indicators are expected to be provided to anticipate the trophic condition of the TGR, and to prevent the occurrence of water bloom based on previously effective measurements. The lack of functional bacteria data from the TGR is particularly worrying because environmental degradation in the TGR will probably be the most severe problem in the future. The aim of this Chapter is to investigate the distributions of total bacteria, culturable bacteria and functional bacteria associated with N or P in the backwater area of the Yangtze River in the TGR, and to further explore a potential indictor reflecting the trophic condition in the TGR.

7.2 Relationship Between Culturable Becteria and Eutropication in the Waterbody

7.2.1 Eutropltication Level

The eutrophication levels of the sampling locations in the Yangzi River were investigated by analyzing their physicochemical characteristics, and it was shown that the eutrophication condition in the studied location was that XX>XJ>DNs (Section 4.2.1).

7.2.2 Culturable Bacteria and Total Bacteria in the Waterbody

The numbers of total bacteria (counted by the acridine orange direct count method, AODC) and culturable heterotrophic bacteria grown in various nutrient levels of a medium, including eutrophic-medium (BEP medium), rich-nutrient medium (PTYG medium), low-nutrient medium (20% PTYG medium) and oligo-nutrient medium (R2A medium) were investigated and the results are shown in Fig. 4.1. To explore the relationship between the bacteria number and the eutrophic condition, the culturable-total bacteria ratios in water and sediment were analyzed (Fig. 7.1; Wang et al., 2010). The culturable-total bacteria ratios in surface water indicated that the ratio values in the investigated rivers were XX>XJ>DN4~N5>DN2~N1>DN3, which was consistent with the sequence of water eutrophic degrees obtained according to the physical-chemical characteristics. It shows that the culturable-total bacteria ratios in the eutrophic waterbody will increase when the pollution in the waterbody becomes more serious. Inversely, the culturable heterotrophicy bacteria are relatively less in the waterbody with a lower eutrophic degree. The culturable-total bacteria ratios based on results using BEP medium and 20% PTYG medium also showed the same trend, namely XX>XJ>DNs, although the sequence of ratios among DNs was not exactly the same as the result achieved from that in the PTYG medium. The sequences of culturable-total bacteria ratios in bottom water and sediment were XX>XJ>DN4~N5>DN2>DN3~N1 and XX>XJ>DN4~DN5>DN3> DN2~N1 respectively, which were similar with those in surface water.

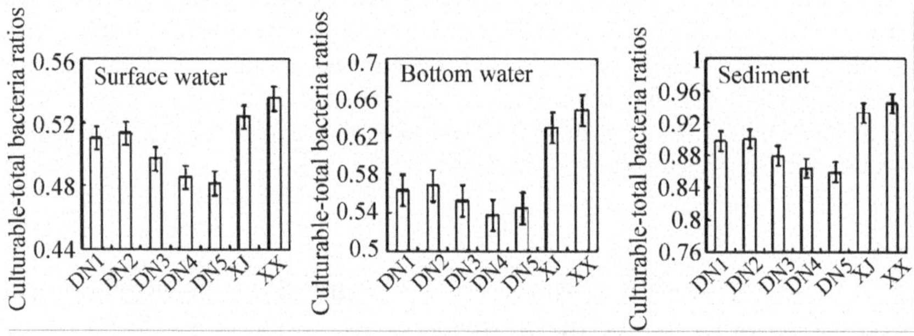

Fig. 7.1 Ratio of culturable heterotrophic bacteria counts to total counts

7.3 Relationship Between Microbial Community and Eutrophication in the Waterbody

Carbon source utilization patterns of microbial communities, also called community- level physiological profiles (CLPP), were assessed using Biolog ECO plates^{TM} (BIOLOG, Inc., Hayward, CA, USA) that contained 31 different carbon sources and a blank in triplicate. Microbial suspensions were prepared with 5 g (dry weight) of sediments in 95 ml of sterile PBS buffer and were shaken for 30 min at 200 r/min on a reciprocal shaker. After settling for 30 min, a volume of 2 mL aliquot of the suspension was diluted in 98 ml of inoculating solution for a final 1:1000 dilution. 100 μL of the sediment suspensions per well were added into the Biolog ECO plates. All solutions, transfer equipment and glassware were sterilized with an autoclave prior to use. Three samples were tested per location, and one ECO plate was used per sample. ECO plates were inoculated at 25 °C in the dark. We plotted corrected color development of the entire plate versus read time to select the optimal periods for the analysis of each plate type (data not shown). Bacteria growth was vigorous in a majority of the plates after 72 h, thus a 72-h incubation period was chosen in the evaluation of bacterial plates. This incubation period was similar to that used in other CLPP studies at a comparable incubation temperature (Classen et al., 2003). Plates were read using an ELISA plate reader (Bio-Rad, Richmond, USA) at 590 nm. The 72-h data were selected for statistical analysis. The average well color development (AWCD) of all 31 carbon sources for each sample was calculated prior to statistical analysis to eliminate variation in well color development caused by different cell densities (Garland, 1996; Liu et al., 2007). All of the optical density (OD) readings were adjusted by subtracting the OD reading in well 1 (water control). Negative values were set to zero.

AWCD is an important factor demonstrating the sole carbon source utilizing capacity of microbes in water or sediments. Fig. 7.2 indicates that the activity of microbes increased with the prolonging of incubation time, and the sole carbon source utilizing capacity of microbes at the same level (surface water, bottom water or sediments) and different sampling sites are XX>XJ>DN1. It also can be seen that the carbon source utilizing capacity is different among different levels in the same sampling site: sediment>bottom water>surface water. It should be noticed that the microbes in the place where there is a higher trophic level have more activity.

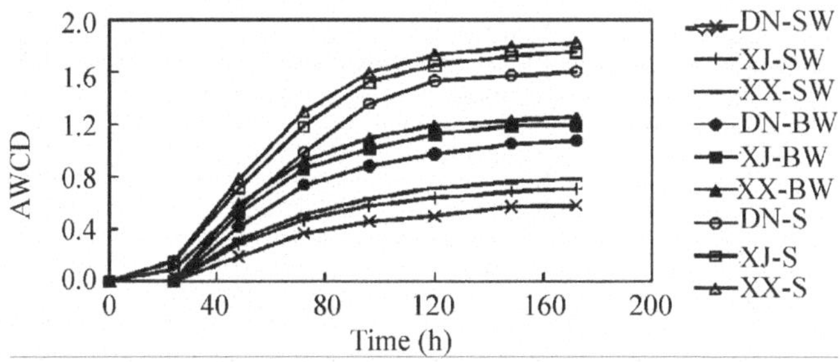

Fig. 7.2 Average well color development (AWCD) of microbial community (DN, Daning River; XJ, Xiaojiang River; XX, Xiangxi River; SW, surface water; BW, bottom water; S, sediment)

The microbial community diversities were investigated using CLPP studies, and the Shannon, Simpson and McIntosh indices were shown in Table 7.4. The three kinds of indices actually reflect different profiles of microbial community diversity. Maguran (1988) pointed out that the Shannon index was mainly effected by the richness of the population species, and Simpson indices mostly reflected the diversities of common species in the population. Atlas (1984) deemed that the McIntosh index could more effectively indicte the homogeneity of population species.

Table 7.4 shows that the diversity indices of the surface water exhibited differences among the three branch rivers (*t*-test, $p>0.05$), in which Simpson and McIntosh indices were significantly different from each other (*t*-test, $p<0.01$). The diversity indices don't significantly differ among the DNs' sites. However, the Simpson and McIntosh indices showed that there exist significant differences between site D1-3 and site D4 and 5 (*t*-test, p<0.01). There also exist significant differences among the bottom water and sediment in DN, XJ and XX (*t*-test, p<0.01). And the diversity indices show that the microbial community diversity level in bottom water is DN1~DN2~DN3~DN4~DN5>XJ>XX, and the diversity levels in sediments is DN1~DN2>DN3>DN4~DN5>XJ>XX.

It can be found that the microbial community diversities in XJ and XX were lower than those in DNs, in opposition to the results of bacteria richness. This was in accordance with the previous reports showing that the increasing nutrient loadings will increase the abundance of bacteria and decrease their diversities (Telesh, 2004).

Table 7.4 Diversity in dices of microbial communities

Sample		Shannon index	Simpson index	McIntosh index
Surface water	DN1	3.59±0.51	34.86±1.19	10.89±1.23
	DN2	3.53±0.23	33.94±1.26	10.61±1.12
	DN3	3.57±0.35	34.25±5.23	10.70±1.22
	DN4	3.43±0.71	30.86±3.28	9.64±1.43
	DN5	3.47±0.47	31.54±2.43	9.86±1.55
	XJ	3.16±0.19	23.47±1.62	7.33±0.68
	XX	3.07±0.62	21.63±1.23	6.76±0.94
Bottom water	DN1	3.97±0.26	43.86±3.53	9.14±0.79
	DN2	3.93±0.34	43.25±2.93	9.01±1.12
	DN3	3.90±0.71	43.14±3.16	8.99±0.95
	DN4	3.86±0.61	42.95±6.23	8.95±0.89
	DN5	3.85±0.23	42.82±5.49	8.92±0.68
	XJ	3.16±0.42	22.34±4.16	4.65±0.57
	XX	3.13±0.19	21.57±3.95	4.49±0.49
Sediment	DN1	4.11±0.21	48.74±5.23	9.20±1.35
	DN2	4.09±0.29	47.43±6.41	8.95±0.96
	DN3	3.96±0.38	43.62±5.63	8.23±1.20
	DN4	3.84±0.49	40.56±4.94	7.65±0.86
	DN5	3.81±0.51	40.12±3.97	7.57±1.28
	XJ	3.55±0.47	31.24±5.13	5.89±0.68
	XX	3.41±0.24	28.18±2.83	5.32±0.76

[a] Average±SD

To distinguish the construction of microbial communities among locations, clustering was performed using Ward's clustering method. Fig. 7.3 (Wang et al., 2010) shows that the seven locations were divided into 2 sets based on the CLPP studies, with the first group including five locations, which were all located in Daning River. The second group included XJ and XX. Qualitative examination of the dendrograms from cluster analysis suggested that the five locations in Daning River had similar bacteria community construction. However, they were largely different from the Xiaojiang River and the Xiangxi River locations.

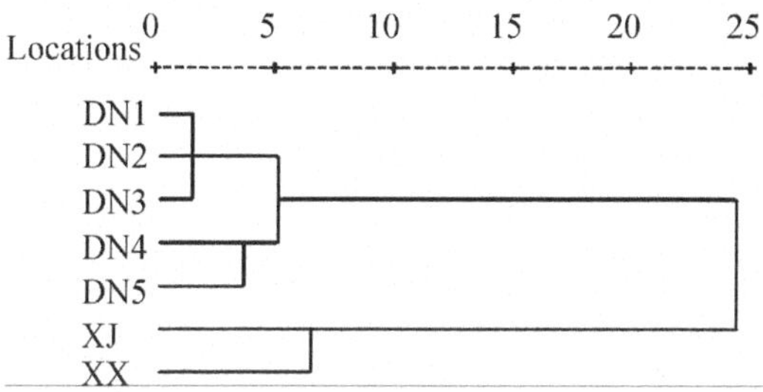

Fig. 7.3 Cluster analysis of diversities of microbial community based on CLPP studies

Principal Component Analysis (PCA) can change the multi-component vectors into disconnected principal elements (PC1 and PC2, the two components of the pivot vector), and the point locations in the pivot vector space after reducing the dimension can directly reflect the metabolic characteristics of different microbial communities. Fig. 7.4 shows the PCA results of the utilization of the carbon source of the microbial community in sediments. The difference in microbial community diversity between the Daning River, Xiaojiang River and Xiangxi River was obvious, whereas the diversity difference among sampling locations in Daning River was small. That was similar to the results from the Shannon index.

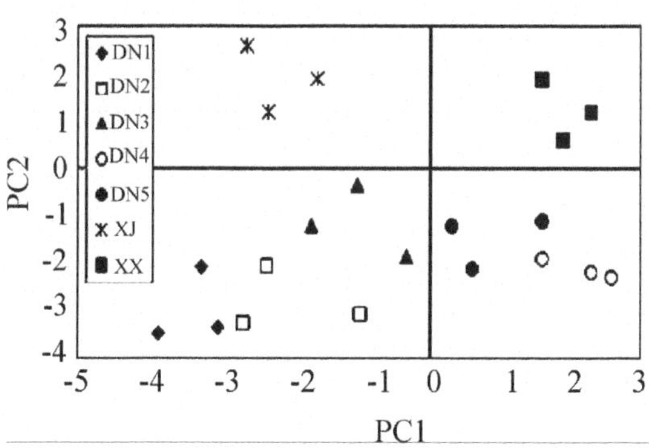

Fig. 7.4 Principal component analysis (PCA) of utilization of carbon source of microbial community in sediments

7.4 Abundance of Functional Bacteria in Aquatic Environments

In order to investigate the distributions of functional bacteria in aquatic environments, eight kinds of functional bacteria, namely AB, NFB, AOB, NOB, NRB, DNB, OPB and IPB in surface water, bottom water and sediments were counted separately. Table 7.5 shows the average, maximum, minimum number of cells and coefficient of variation (CV) in surface water, bottom water and sediments from the seven locations (Wang et al., 2010). It shows that the CVs of functional bacteria in water and sediments were high, which indicated that the variation in bacterial richness among the seven locations were high. This predicted that the change in the aquatic environment could induce a change in bacterial distribution in the aquatic ecosystem.

Table 7.5 Counts of functional bacteria associated with N or P (log of cell number ml^{-1} or g^{-1})

Sample	Surface water				Bottom water				Sediments			
	Ave	CV%	Max	Min	Ave	CV%	Max	Min	Ave	CV%	Max	Min
AB	3.50×10^5	30.1	5.22×10^5	1.93×10^5	1.72×10^5	35.5	2.75×10^5	1.04×10^5	6.86×10^8	29.6	1.09×10^9	5.75×10^8
NFB	2.44×10^5	146.2	9.44×10^5	3.60×10^4	8.83×10^5	77.7	2.24×10^6	4.21×10^5	1.48×10^7	94.6	4.60×10^7	8.40×10^6
AOB	2.22×10^2	85.2	5.50×10^2	1.06×10^2	5.10×10^4	89.2	1.30×10^5	1.95×10^4	2.40×10^5	28.2	3.50×10^5	1.60×10^5
NOB	4.49×10^2	74.6	1.10×10^3	1.80×10^2	7.16×10^5	64.6	2.60×10^6	3.10×10^5	3.48×10^7	179.7	7.10×10^7	6.20×10^6
NRB	1.45×10^5	35.6	2.00×10^5	6.00×10^4	1.40×10^5	59.4	2.45×10^5	3.00×10^4	3.78×10^7	73.0	9.44×10^7	1.08×10^7
DNB	7.63	32.0	12.00	4.00	66.00	100.7	200.00	21.20	8.75×10^5	147.1	2.90×10^6	7.28×10^4
OPB	1.43×10^3	116.8	4.67×10^3	2.61×10^2	3.69×10^3	133.0	1.26×10^4	7.12×10^2	1.37×10^5	83.8	3.83×10^5	4.57×10^4
IPB	1.51×10^4	38.1	2.53×10^4	8.30×10^3	2.24×10^6	67.9	5.64×10^6	1.30×10^6	1.97×10^7	105.5	6.17×10^7	8.18×10^6

AB, ammonifying bacteria; NFB, nitrogen-fixing bacteria; AOB, ammonia-oxidizing bacteria; NOB, nitrite-oxidizing bacteria; NRB, nitrate-reducing bacteria; DNB, denitrifying bacteria; OPB, organophosphate-solubilizing bacteria; IPB, inorganic phosphate-solubilizing bacteria; CV, coefficient of variability

To better demonstrate the distribution characteristics of these functional bacteria, the distribution ratios of surface water to bottom water and ratios of water to sediments were calculated and shown in Fig. 7.5 (Wang et al., 2010). The data used to calculate these ratios were the logarithm values of the colony count. Since the amount of bacteria in sediments was relatively stable, these ratios could reflect the change in functional bacteria richness in the water to a certain degree. Furthermore, these ratios could indicate the nutrient conditions of the water according to the functional bacteria richness in the water. For example, the NOB ratios of water to sediments were 0.827, 0.816, 0.845, 0.815, 0.832, 0.915 and 0.923 for DN1, DN2, DN3, DN4, DN5, XJ and XX, respectively. This showed that NOB ratios in XJ and XX were obviously higher than those in DNs, which basically was consistent with the nutrient loadings in these seven locations. Fig. 7.5 shows that for different kinds of bacteria their distribution ratios, each limited to a certain range, were different from each other. The ratios of AB, DNB, NRB and IPB exhibited weak difference among the seven locations, which indicated that these four kinds of functional bacteria were not very sensitive to the change of environment. The ratios of NFB, AOB, NOB and OPB in DNs locations were obviously different from those in XJ and XX locations, indicating that ratios of these four kinds of bacteria exhibited different values in diverse aquatic environments. In other words, the change of ratios of NFB, AOB, NOB and OPB could reflect the change in the trophic condition of the aquatic environment.

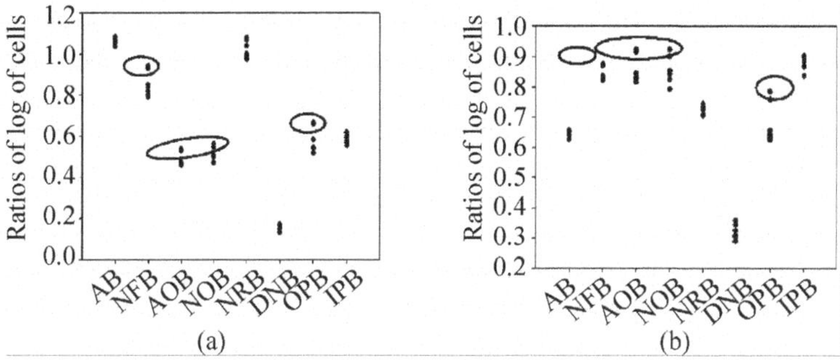

Fig. 7.5 Distribution ratios of functional bacteria in different aquatic ecosystems. (a) Ratios of log of cells between water and sediments; (b) Ratios of log of cells between surface and bottom water

Since the habitats in sediments were relatively stable, the bacteria ratios of water to sediments were more sensitive to the change of aquatic ecosystem. The data of these ratios were used to group the aquatic environment using cluster analysis. Fig. 7.6 (Wang et al., 2010) indicated that the seven locations were divided into 2 sets, which was similar to the grouping based on the microbial community construction. This indicated that, like the microbial community construction, the distribution of functional bacteria richness in aquatic environments could also reflect the water quality condition. This provided a cost effective and intuitionistic method to indicate the trophic condition of the water.

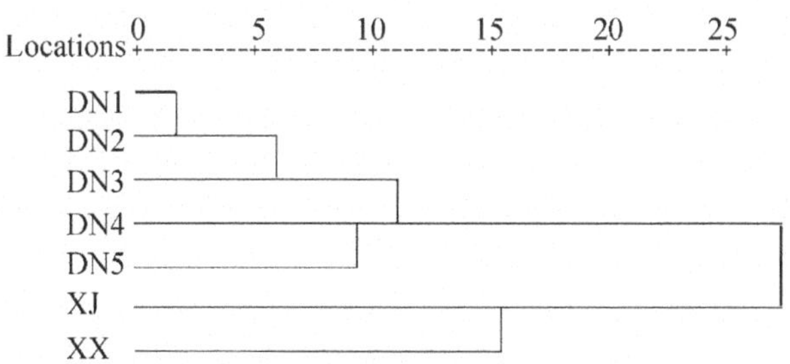

Fig. 7.6 Cluster analysis of microbial community construction based on the distribution ratios of functional bacteria in aquatic environments

7.5 Eutrophication Risk Assessment and Hydraulic Control in Large Reservoirs

7.5.1 Sensitivity Evaluation for Eutrophication Risk in Large Reservoirs

Many methods, such as the characteristic method, parameters method, biological index evaluation method, phosphorus income and expenses model, nutrition state index method, mathematical analysis method, GIS basis evaluation methods, etc. are used to assess the eutrophication risk in large lakes and reservoirs (Liu, 2011; Cai, 1993). Here, combined with the simulation results from macro-scale one-dimensional water quality modeling in Section 6.3, the scoring method was used in assessment eutrophication risk sensitivity in the Three Gorges Reservoir Area.

Three levels, poor eutrophication, intermediate eutrophication and rich eutrophication, were adopted in the scoring method. Each eutrophication level will have several sub-levels for detailed division. Choose a certain number in the index as the basis of water body eutrophication score factors, each index will be scored as 0–100 points and a linear interpolated method will be used when the scoring value is between sub-levels. The higher the score value is, the more serious the water eutrophication level will be. In this chapter, we select four eutrophication key parameters. They are TN, TP, COD_{mn} and velocity of flow (V) as the evaluation index. The assessment criteria adopted was that of Zhong (2000) (Table 7.6). The evaluation equation is as follows:

$$M = \sum_{i=1}^{n} M_i / n \tag{7.1}$$

where M is the eutrophication assessment value, M_i is scoring value of the evaluation index, n is the number of the evaluation index.

The four indexes in the dry, normal and rich season in the Three Gorges Reservoir Area are from the simulation results of macro-scale 1-D water quality modeling in Section 6.3. The above scoring method was used; the assessment results are as follows.

In the dry season, the river sections with higher eutrophication assessment value including: Chongqing to Zhongxian, Wanzhou to Yunyang, Fengjie to Wushan, Xiangxi River mouth to Three Gorge Reservoir dam, and the blackwater zones of tributaries (Fig. 7.7). The backwater zones of the tributaries are the area with highest eutrophication assessment value. The reason for this is that there is less water quantity coming from upstream, a high water level before the dam and slow water velocity in the dry season. Especially in March and April each year, with appropriate temperature and light, algal bloom will occur frequently in the backwater zones of tributaries. There is no report about algal bloom in the mainstream so far in this area.

Table 7.6 Assessment criteria of eutrophication status at the TGR

Scoring	Eutrophication type	Index			
		TP (mg/L)	TN (mg/L)	V (m/s)	COD_{mn} (mg/L)
0	Extremely poor	<0.0025	<0.025	>3.5	<0.2
10	Poor	0.005	0.05	2	0.4
20	Poor	0.0075	0.075	1	0.8
30	Poor-intermediate	0.01	0.1	0.5	1.5
40	Intermediate	0.025	0.3	0.3	3.0
50	Intermediate-rich	0.05	0.6	0.10	4.0
60	Rich	0.10	1.2	0.05	7.0
70	Seriously rich	0.15	2.7	0.01	10
80	Extremely seriously rich	0.25	4.0	0.005	20
90		0.40	5.0	0.002	40
100	Particularly seriously rich	>0.50	>6.0	<0.001	>60

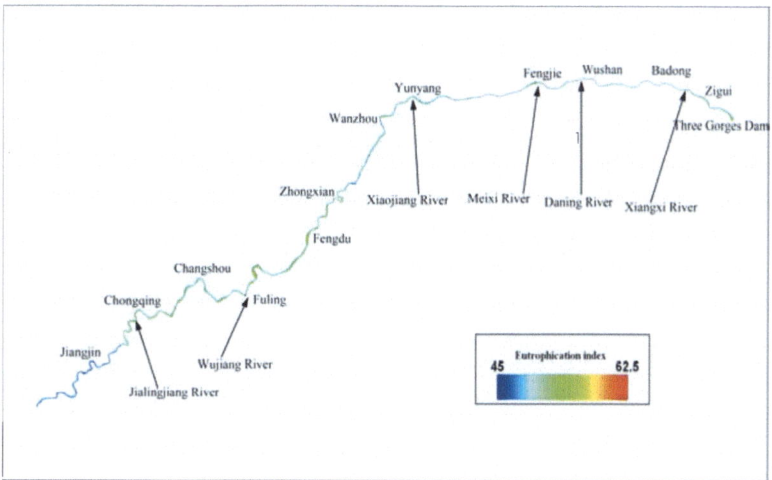

Fig. 7.7 Eutrophication assessment value during dry season in the TGRA

In a normal season, the river sections with higher eutrophication assessment value include Chongqing to Zhongxian and the backwater zones of tributaries. The eutrophication assessment value in this area is between 42.5 and 65. At this time, the water level before the dam still remains high. There is more water from upstream than in the dry period which carries a certain amount of TP into the reservoir, together with water.

Temperature and light conditions, occurrence probability of algal bloom in river courses backwater zones is great.

In the rich season, there is high water quantity from upstream. For example, the peak flow on July 20, 2010, reached 70,000 m^3/s. At the same time, for the purpose of flood control, the water level before the dam was 145 m, the fall in water level between the river upstream and downstream is big and the water velocity is large. In addition, a large amount of sediment carried by the flow makes the water become turbid which prevents algal growth by photosynthesis. But in some backwater zones, such as river mouth of the Daning River, Xiangxi River, the water flow velocity in the mainstream has less influence on these regions, and a high concentration of TP in water might provide good nutrition conditions. Also, the eutrophication assessment value is high in these areas.

7.5.2 Hydraulic Control Technology for Prevention of Eutrophication in Large Reservoirs

The operation and management of the Three Gorges Reservoir Project is a multi-objective decision-making issue. The following targets should be included: flood control, power generation, navigation, ecological environment protection, and so on. In the current study, the ecological environment protection target is mainly concerned with fish breeding conditions (Cao and Cai, 2008), but environmental water management and protection is mentioned less. Fig. 7.8 shows the operation curve of the Three Gorges Project.

Since algal bloom has become the main issue relating to water quality in the Three Gorges Reservoir Area, and there is a close relationship between algal bloom and water velocity, by adjusting the water level before the dam to change the water flow velocity, the occurrence probability of algal bloom may be reduced.

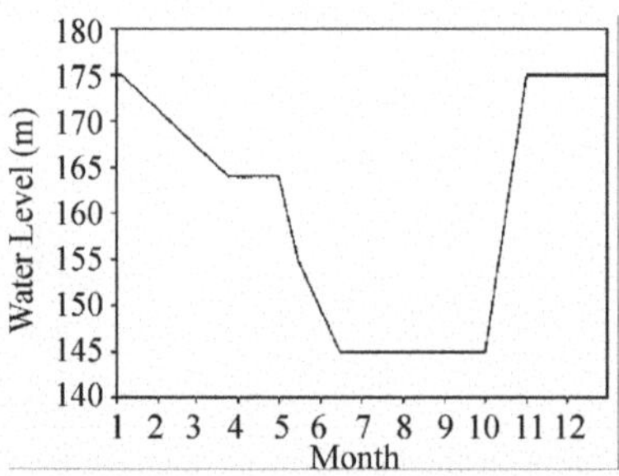

Fig. 7.8 Operation curve of the Three Gorges Project

In this chapter, we assessed the effect of water level control on eutrophication by scenario planning. We assumed a running model of 175 m, the level decreasing by 0.5 m/h, 0.5 m/3 h, 0.5 m/6 h, 0.5 m/12 h, 0.5 m/24 h, 5 m/d, 5 m/3 d, 5 m/5 d before the dam. The water flow velocity and eutrophication index can be achieved by running a macro-scale one-dimensional integrated model in Section 6.3. Then the change in eutrophication assessment value at the four main river mouths, Xiangxi, Daning, Meixi and Xiaojiang, can be obtained. It is assumed that the water level operation occurred on the 73rd day after New Year's Day, which corresponds to 15th March. Actually it is the earliest day that eutrophication occurred during these years in the Three Gorges Reservoir Area.

Some results in difference scenarios can be seen from Figs. 7.9–7.16. These figures show that in a scenario of 0.5 m/h, the river mouths near the dam, such as the Xiangxi River mouth and Daning River mouth have a certainty impact, and for a scenario of 5 m/day, all the four river mouths have a certainty impact. Other scenarios have little impact on the prevention of eutrophication. So much research needs to be done on understanding hydraulic control technology so as to prevent eutrophication.

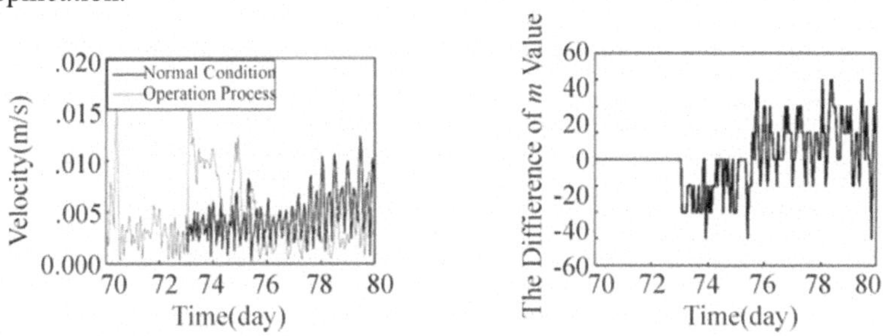

Fig. 7.9 Velocity (*L*) and eutrophication assessment value (*R*) in Xiangxi River mouth (0.5 m/h)

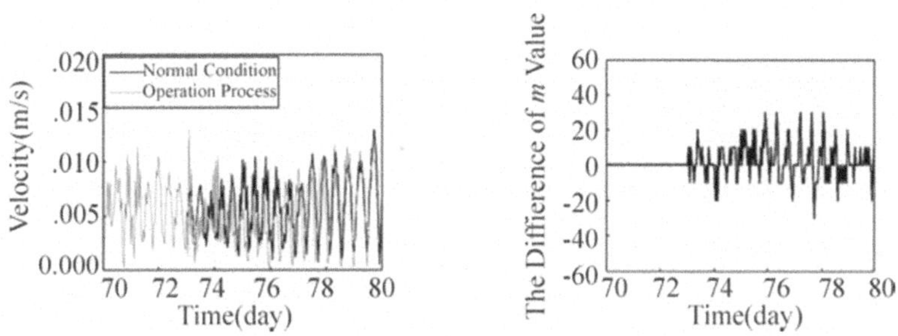

Fig. 7.10 Velocity (*L*) and eutrophication assessment value (*R*) in Xiangxi River mouth (0.5 m/3 h)

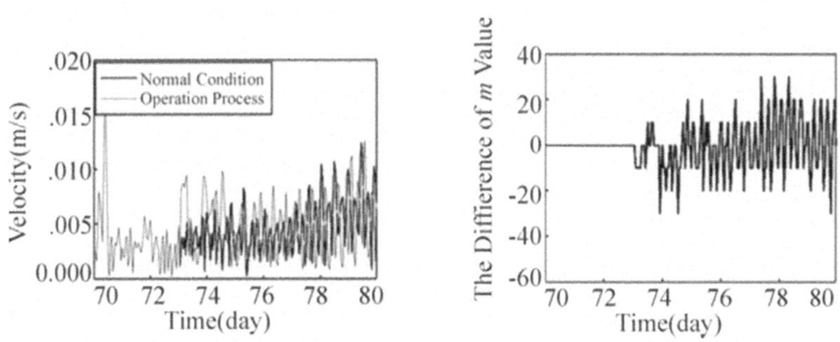

Fig. 7.11 Velocity (*L*) and eutrophication assessment value (*R*) in Daning River mouth (0.5 m/6 h)

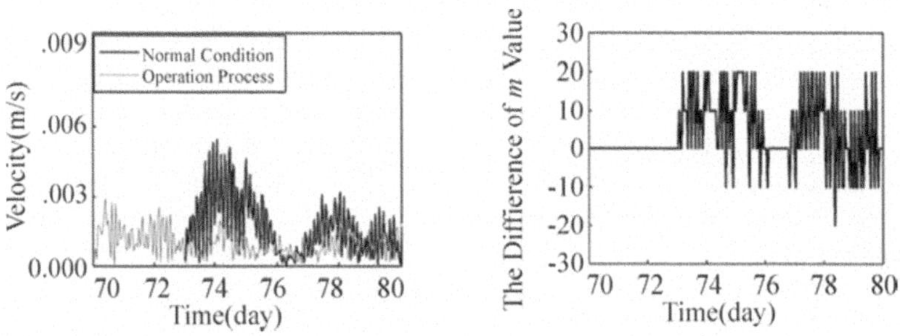

Fig. 7.12 Velocity (L) and eutrophication assessment value (R) in Meixi River mouth (0.5 m/12 h)

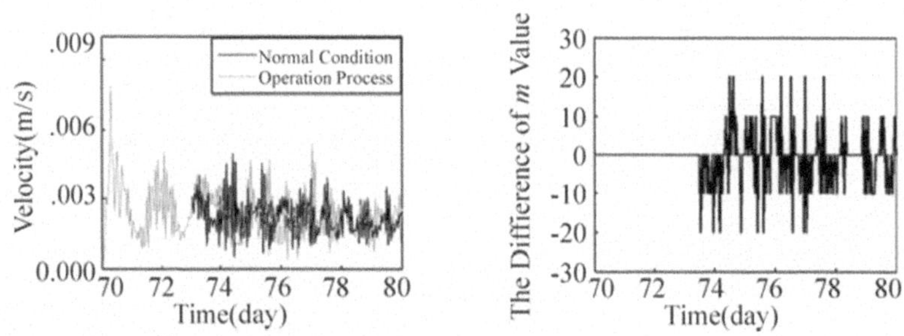

Fig. 7.13 Velocity (L) and eutrophication assessment value (R) in Xiaojiang River mouth (0.5 m/24 h)

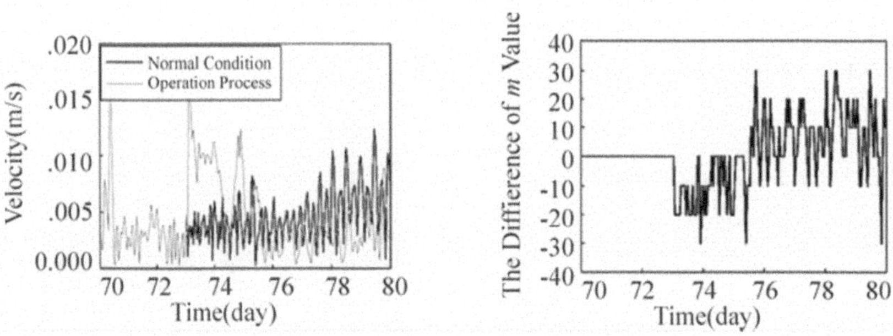

Fig. 7.14 Velocity (L) and eutrophication assessment value (R) in Daning River mouth (5 m/d)

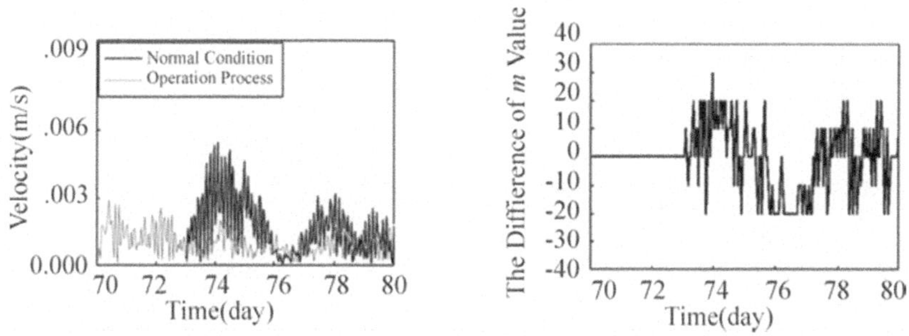

Fig. 7.15 Velocity (*L*) and eutrophication assessment value (*R*) in Meixin River mouth (5 m/3 d)

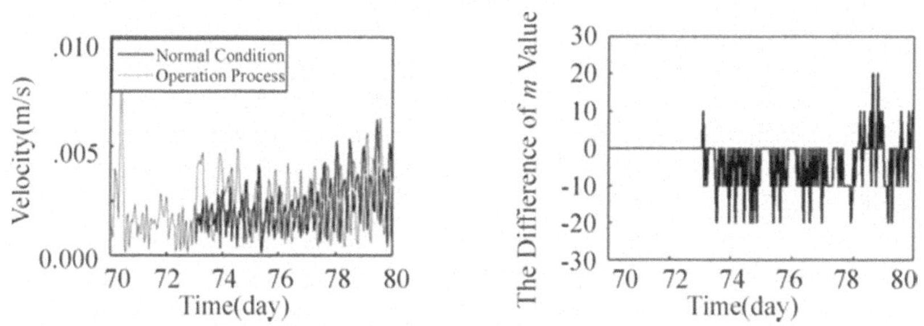

Fig. 7.16 Velocity (*L*) and eutrophication assessment value (*R*) in Xiaojiang River mouth (5 m/5 d)

The aquatic ecosystems of the backwater areas of the Yangtze River in the TGR, where water bloom frequently occurred, were in an eutrophic condition. Eutrophic conditions in the Xiaojiang River and Xiangxi River were more severe than in the Daning River. The total bacteria and culturable bacteria were relatively abundant in the TGR. Our experimental results revealed that the richer the nutrient level of media was, the greater were the culturable bacterial colonies cultured in eutrophic freshwater. The results of bacteria counting and CLPP studies revealed that the increase in nutrient loadings would give rise to bacteria densities yet decrease bacteria community diversities.

Eight kinds of functional bacteria associated with N or P, namely AB, NFB, AOB, NOB, NRB, DNB, OPB and IPB, were found to be abundant in the backwater areas in the TGR. Among the eight kinds of functional bacteria, four kinds of bacteria (NFB, AOB, NOB, OPB) displayed sensitivity to changes in the trophic conditions of the water. It has been observed that the location grouping result from cluster analysis, based on the functional bacteria ratios of water to sediments, was similar to that based on the CLPP studies. The ratios of functional bacteria could distinguish the aquatic environments with different trophic conditions in the TGR. Thus, the functional bacteria ratios of surface water to bottom water or ratios of water to sediments may be used as potential indicators of water quality in the TGR.

References

Borja A (2004) The biotic indices and the Water Framework Directive: the required consensus in the new benthic monitoring tools. Mar Pollut Bull 48:405-408.

Cai QH (1993) Eutrophication comprehensive evaluation of East Lake in Wuhan. Oceanologia & Limnologia Sinica 24:335-339.

Cao GJ, Cai ZG (2008) Integrated operation and management of the Three Gorges Reservoir: A review. The Three Gorges Reservoir Construction 3:41-44.

Celussi M, Cataletto B (2007) Annual dynamics of bacterioplankton assemblages in the Gulf of Trieste (Northern Adriatic Sea). Gene 406:113-123.

Chainho P, Costa JL, Chaves ML, Dauer DM, Costa MJ (2007) Influence of seasonal variability in benthic invertebrate community structure on the use of biotic indices to assess the ecological status of a Portuguese estuary. Mar Pollut Bull 54:1586-1597.

Classen TA, Boyle IS, Haskins EK, Overby TS, Hart CS (2003) Community-level physiological profiles of bacteria and fungi: plate type and incubation temperature influences on contrasting soils. FEMS Microbiol Ecol 44:319-328.

Davidson K, Gilpin LC, Hart MC, Fouilland E, Mitchell E, Calleja IA, Laurent C, Miller AE, Leakey RJ (2007) The influence of the balance of inorganic and organic nitrogen on the trophic dynamics of microbial food webs. Limnol Oceanogr 52:2147-2163.

Garland JL (1996) Analytical approaches to the characterization of samples of microbial communities using patterns of potential C source utilization. Soil Biol Biochem 28:213-221.

Hwang SS, Xi J, Cao Y, Feng X, Qiao X (2007) Anticipation of migration and psychological stress and the Three Gorges Dam project, China. Soc Sci Med 65:1012-1024.

Jergensen BB, Richardson K (1996) Eutrophication of Coastal Marine Systems. American Geophysical Union, Washington, DC.

Kim LH, Choi E, Stenstrom MK (2003) Sediment characteristics, phosphorus types and phosphorus release rates between river and lake sediments. Chemosphere 50:53-61.

Liu B, Tu C, Hu S, Gumpertz M, Ristaino JB (2007) Effect of organic, sustainable, and conventional management strategies in grower fields on soil physical, chemical, and biological factors and the incidence of southern blight. Appl Soil Ecol 37:202-214.

Liu HL (2011) Lake Eutrophication Control. China Environment Science Press, Beijing.

Nedwell DB, Sage AS, Underwood GJ (2002) Rapid assessment of macro algal cover on intertidal sediments in a nutrified estuary. The Science of The Total Environment 285:97-105.

Nixon SW (1995) Coastal marine eutrophication: a definition, social causes and future concerns. Ophelia 41:199-219.

Paerl HW (1997) Coastal eutrophication and harmful algal blooms: Importance of atmospheric deposition and groundwater as new nitrogen and other nutrient sources. Limnol Oceanogr 42:1154-1165.

Potapova M, Charles DF (2007) Diatom metrics for monitoring eutrophication in rivers of the United States. Ecological Indicators 7:48-70.

Simboura N, Zenetos A (2002) Benthic indicators to use in ecological quality classification of Mediterranean soft bottom marine ecosystems, including a new biotic index. Mediterr Mar Sci 3:77-111.

Telesh IV (2004) Plankton of the Baltic estuarine ecosystems with emphasis on Neva Estuary: a review of present knowledge and research perspectives. Mar Pollut Bull 49:206-219.

Wang HY, Shen ZY, Guo XJ, Niu JF, Kang B (2010) Ammonia adsorption and nitritation in sediments resourced from Three Gorges Reservoir. China Environ Geol 60:1653-1660.

Zboinska E, Maliszewska I, Lejczak B, Kafarski P (1992) Degradation of organophosphonates by Penicillum citrinum. Lett Appl Microbiol 15:269-272.

Zhong CH (2000) Eutrophication in Three Gorges Reservoir Area. Sichuan University Press, Chengdu.

Index

The manufacturer's authorised representative in the EU is Springer Nature Customer Service Centre GmbH, Europaplatz 3, 69115 Heidelberg, Germany. If you have any concerns regarding our products, please contact ProductSafety@springernature.com

Printed and bound by CPI Group (UK) Ltd, Croydon, CR0 4YY
12/08/2026
02194517-0002